WHO IS THE EARTH?

HOW TO SEE GOD
IN THE NATURAL WORLD

BY THE SAME AUTHOR

Doorkeeper of the Heart: Versions of Rabi'a

Hammering Hot Iron:
A Spiritual Critique of Robert Bly's Iron John

The System of Antichrist:
Truth and Falsehood in Postmodernism and the New Age

Cracks in the Great Wall:
The UFO Phenomenon and Traditional Metaphysics

Legends of the End:
Prophecies of the End Times, Antichrist,
Apocalypse, and Messiah from Eight Religious Traditions

The Virtues of the Prophet:
A Young Muslim's Guide to the Greater Jihad,
The War Against the Passions

Knowings: In the Arts
of Metaphysics, Cosmology, and the Spiritual Path

Folk Metaphysics:
Mystical Meanings in Traditional Folk Songs and Spirituals

Reflections of Tasawwuf
Essays, Poems, and Narrative on Sufi Themes

Shadow of the Rose:
The Esoterism of the Romantic Tradition
(with Jennifer Doane Upton)

The Wars of Love

CHARLES UPTON

WHO IS THE EARTH?

HOW TO SEE GOD IN THE NATURAL WORLD

INCLUDING THE ESSAY:

ATLANTIS AND HYPERBOREA:
AN ENQUIRY INTO THE
CYCLICAL MYSTERIES

SOPHIA PERENNIS

SAN RAFAEL, CA

First published in the USA
by Sophia Perennis
© Charles Upton 2008

Series editor: James R. Wetmore

For information, address:
Sophia Perennis, P.O. Box 151011
San Rafael, CA 94915
sophiaperennis.com

Library of Congress Cataloging-in-Publication Data

Upton, Charles, 1948–
Who is the Earth?: how to see God in the natural world,
including the essay 'Atlantis and Hyperborea: an enquiry
into the cyclical mysteries' / Charles Upton.

p. cm.
ISBN 978-1-59731-072-7 (pbk: alk. paper)
1. Nature—Religious aspects. 2. Spiritual life.
3. Cosmology. I. Title
BL65.N35U68   2008
202'.4—dc22   2008003233

CONTENTS

PART TWO:
ATLANTIS AND HYPERBOREA

PREFACE

WE DESTROY NATURE because we don't really see it. We don't see it because we don't know what it really is. It's not only that most of us now live in cities; plenty of people living in rural areas destroy the natural world around them as a matter of course, hardly realizing what they've done. This is because our culture possesses no unified vision of the nature of what we call 'nature'. We see it as a collection of obstacles; as a mass of exploitable resources; as a set of interlocking mechanical processes — or, alternately, as a setting for leisure, an opportunity for aesthetic enjoyment, a magic world of subtle energies, or a Great Goddess with her retinue of spiritual forces. All of these visions have a degree of truth to them, some much more than others. But the one vision we find hardest to maintain is the truly unified one—that of the natural world as a living symbol of its Creator.

This book is an attempt to say what nature is, and what it is not. Herein I critique some of the common myths of nature, and re-present the natural world not as an independent deity, but as the primordial symbol of God. In order to see the Earth like this, we must 'cleanse the doors of perception'. But we also need to sense how, as we contemplate the Universe around us, we are also being contemplated; as we sit, quietly watching the Earth, Someone Else is watching us. And we need to realize (paraphrasing Meister Eckhart) that the eye through which we see the Earth, and the eye through which that Someone sees us, is the same eye.

Something that we can only look at is a thing, an object, an 'it'. Something that, while we are watching it, is also watching us, is a being, a someone, a 'who'. If, while contemplating the natural world around us, we can understand intuitively exactly Who is contemplating us, then we will know the answer to the question that is the title of this book: 'Who is the Earth?'

PART ONE

SEEING GOD
IN THE NATURAL WORLD

Where man is not,
nature is barren.

—WILLIAM BLAKE

I

WHAT IS 'NATURE'?

I IMAGINE THAT MOST OF US, if we ever ask this question, tend to answer it something like this: 'Nature is a beautiful landscape of trees, plants and wildlife, sometimes including lakes, rivers, and high mountains, or desert scenes with interesting rock formations—oh yes, and the Ocean.' That's the sightseer's answer, the photographer's or painter's answer.

Those whose minds are a bit more analytical may define Nature as 'anything material that isn't us, or what we build, or what we destroy'.

The technologist answers that Nature is 'the sum total of resources, minerals, plants and animals, that are useful to human beings.'

Some farmers, most architects, and all builders of highways think of Nature as 'that mass of physical substance that supports or gets in the way of what I'm trying to get done.'

Dante, following Aristotle, says that Nature is God's artwork, the model for all human art and craft.

The philosophers of the Enlightenment thought of Nature as the source of human 'equality'. Society, being artificial and contrived, makes us unequal, it divides us into classes, but Nature is an impartial mother who gives freely and equally to all. (How true is this, really?)

The Muslims and the Church Fathers say that Virgin Nature is God's primordial scripture, His first book.

The scientist defines Nature as 'that which responds, via measurable physical data, to the questions we ask it in our physical experiments'.

One of the most interesting answers is the Taoist one, which defines Nature as 'that which arises of itself'.

'What arises of itself' includes a wildflower, an ecosystem, a heartbeat; a feeling, a reverie, a moment of intimacy; a dream, a change in the weather, or the whole universe of stars. It might also include a war, a cultural Renaissance, a spiritual vision, an artistic inspiration, or a trend in the stock market. It encompasses whatever is dynamic, unselfconscious, sometimes predictable, but never controllable, either because it is too far 'without', in the maze of circumstances, or too far 'within', in the secret place of God, to be governed by the human ego. It simply happens, whether or not we are watching, whether or not we are trying to understand. In Judeo-Christian terms it is Providence, the Will of God.

'What arises of itself' comes before the ego and its considered ideas. It's the raw material our ego works on, distorts, 'improves', classifies, denies. But since this ego also 'arises of itself', we can't leave it entirely out of the picture. The trick is to keep it from taking over the whole universe, as an invasive foreign shrub will cover vast areas, so that many native plant species can no longer grow. As Lao Tzu said, 'Do you think you can take over the universe and improve it? I do not believe that it can be done.'

Seen in this way, human civilization and technology are 'natural' too. Civilization is as natural to humanity as hives are to bees. Technology is as natural to humanity as horns are to stags. The trick is, not to let it take over the whole field, like imported Scotch broom crowding out the coyote bush, the native flowers and grasses—a trick we have not yet learned.

Technology in its present gross, destructive form, developed in the modern West, is largely a product of the most negative aspects of the human ego. And while it is true that this ego 'arises of itself', so that it is correct to say that artificiality itself is a natural development, it is also true that, once we are under the power of the ego, we can no longer see this.

So something else is needed, something beyond the 'natural' in the sense of the 'automatic', something based on human intelligence, responsibility, and work—not the sort of work the ego does in its blind automatism, which is the very thing that hides from us all that 'arises of itself', but the sort of work that refines and elevates human consciousness until the darkness of the ego is dispelled, work in which human artificiality takes its cues from what 'arises of itself', and serves it.

For a human being to refine his or her character is as natural as for a cat to groom itself, or for geese to fly in a V-formation. It's just what we do. It arises of itself. But when we forget this—which only we of all 'animals' can do, because only human beings are capable of developing an ego that is at odds with our true nature—when, that is, we deny our instinct for conscious self-development—then the result is as absurd and as catastrophic as if cats were to suddenly all try to fly, or geese to catch mice. It's contrived. It's unnatural.

The struggle against nature is natural to humanity; and so it is up to us to choose whether we will struggle primarily against the natural world by means of the ego, or against the ego itself, by means of the Grace of God. That's the real crux.

⟲

II

THE OTHER SIDE

WHEN WE GO OUT INTO NATURE, into that part of the planet which is neither destroyed nor cultivated by human action, we meet a different part of ourselves—the part that 'arises of itself'.

Is this the 'real me'? Not necessarily. In a way it's more simple, more genuine than the self we cultivate to deal with human society; it isn't quite so contrived. But since our social self is just as real and just as natural to us as our 'natural' self, we can't call our natural self 'realer'; it's just the other side of our being.

A child is not realer than an adult. An acorn is not realer than an oak tree. Direct experience through our senses and subtle intuitions is not realer than well-developed thought and refined feeling, any more than a mountain is realer than a city. But it is certainly just as real, which means we'd better not ignore it.

๑

III

THE PROBLEM
WITH NATURE-WORSHIP

BEFORE the EGO is born, there is no 'nature' because there is no 'me'.

After the EGO is born, there is a 'me in here' peering out of its hole at a 'nature out there'.

So the problem with worshiping nature is the same as the problem with worshipping yourself. There is no breakthrough to a higher level, no liberation; everything remains within the circle of the EGO.

People often say, 'Don't treat me as an object; I'd rather be loved than worshipped.' It's the same with the Earth, in some ways. To worship her is to make her into an object, an idol; it is to relate to her hungrily, possessively, filled with fear and desire—or with sentimentality, or with cheap mythopoetic glamour.

To worship the Earth is to manipulate and exploit her by turning her into the object of one's ego. If we must worship something— and unless we willingly submit to a Reality higher than our EGO, we remain subhuman—then we had better worship GOD, not nature, because GOD is the 'memory' of a 'time' before the EGO was born, when there was no 'nature', when there was no 'me'.

When we obsessively try to re-connect with an object 'out there' called 'nature' by possessing it—subtly, through psychic projection, or grossly, by buying up acreage—we only widen the split. It's like encountering, say, a deer. If you jump up to embrace the beautiful

beautiful deer, she will only run away—and she can run a lot faster than you can. But if you just sit still, and don't care, ultimately, if she approaches or not, then maybe she will approach. And the only way, ultimately, to sit that still, and love that much, without 'caring' at all, is to remember the time before the self and the world were split, and look forward to the time when they will *inevitably* be reunited, and know both times as Now.

Because that time never really ended. It's as if we were still walking on it without knowing it, like the ground under our feet. GOD is the intimate knowledge that that time never ended.

🌀

IV

THE MOTHER/CHILD TRAP: NATURE MUST SAVE ME

MATERIAL NATURE gives us our bodies, and the field of subtle life-energy within which our bodies thrive. That's all she gives us. That's all we have a 'right' to expect. Anything else we get from her is a product of human intelligence and labor—of human society. And, of course, the way things are now set up, and have been for many thousands of years, our bodies are also a product of human society, because the food we eat is a result of social organization; this has been true since the first agriculture, the first organized hunting. But in the most basic sense, underlying all society and all life-support-ing technology, nature gives us our food, our water, our air, our physical life. Only the hypnotizing effect of our recent encapsula-tion within totally artificial environments makes it necessary to repeat this glaringly obvious fact.

So even though *imaginal* Nature can give us a universe of living symbols of our Divine origin, the only thing physical nature gives us is physical life. Very often, however, we expect more. We expect nature to give us love, security, a feeling of wholeness. The love we never got from our human mothers, or never assimilated, we expect to get from Mother Earth.

But Mother Earth cannot give us human love; that's not her role. Nor (as we know) can she really give us security: because nature is ruthless. Nor can she give us wholeness, except by putting us in touch with that 'other side' of our being which is part of that whole-ness, a part we often forget.

People who want the Earth to save them, who want to physically or psychologically return to the womb of the Great Mother, are basically asking for death. They are, in the words of Keats, 'half in love with easeful death.' When we ask the Earth to make us whole, we are projecting our inner image of wholeness on the natural world around us, either as the mother's womb, or as the 'peace of the grave'.

But a foetus is not a whole human being, is it? And a moldering corpse is not a whole human being either. Wholeness is elsewhere. And although the *Bardo Thodol* (the *Tibetan Book of the Dead*) teaches that death gives us a chance at wholeness, a chance to recognize our true nature—the Clear Light of the Void—it also teaches that we can't pick up on that chance simply by dying, or by returning to the womb in rebirth. Our wholeness is in eternity, in God, in the depth of the present moment—but we have to know how to recognize this, through spiritual work, or it won't do us a bit of good.

To project this wholeness on the screen of an external nature is basically a flight from this recognition.

๑

V

WORK

THE POET WILLIAM CARLOS WILLIAMS once wrote:

We cannot go to the country
For the country will bring us no peace.

Is this just the gripe and rationalization of an inveterate urban solid citizen from New Jersey? Possibly. Nonetheless, if what we are looking for in nature is the wrong kind of 'peace', we will never set foot on her true ground. No real family farmer (a few are left) ever says to himself, 'Ah, the peace of nature, the silence of the snowy fields' without remembering his mortgage, his potato blight, and his aching back.

To the Native Americans, Mother and Grandmother Earth (visible nature and underlying divine Substance) was not 'the country'. It was at once workplace, residential area, and temple. It was, and is, rigorous. A fruitful relationship with it requires balance, vigilance, and hard work. The law here is: 'No one can relate to the natural world without physical work and not get drained.'

Our society defines nature as, among other things, 'the setting for leisure'. But leisure, for us, is often far from relaxation: Dirt bikes. Power boats. Snowmobiles. (Not to mention boom-boxes.) Is this because we have to blow off a lot of urban steam, and have only a short time to do it in? Maybe.

But it's also true that a healthy relationship with nature requires effort, which means that those who want to lay back and sip their drink and smoke their joint and indulge themselves in a Beauty without Rigor, a thornless rose, or fondle a stinkless skunk, or wrestle playfully with a de-fanged cougar, will gradually become filled with fluttering moths, and long phosphorescent worms, and various other clots of displaced etheric energy, elf-bubbles and pixie-dust, and lose their human form. Nature is Earth-element, and Earth-element is work.

⟋

One of the benefits of physical labor, especially in nature, is that it teaches us not to fight gravity. If you are walking around inside your head, exclusively identified with your thinking mind, then gravity becomes your enemy; you experience it as bondage, stagnation, dead weight. But if you let your attention fall, naturally, to the center of the Earth (while exhaling), then life energy will naturally rise up, from the center of the Earth, and fill you (as you inhale). As Jesus said: 'He who exalts himself shall be humbled, but whoever humbles himself shall be exalted.'

⟋

VI

THE BEAUTIFUL
AND THE SUBLIME

TRADITIONAL AESTHETICS talks about the Beautiful and the Sublime. The Beautiful brings joy and peace. The Sublime inspires awe and terror.

The Beautiful and the Sublime are the two essential qualities of the natural world: The wooded lake and the erupting volcano; the peacock and the cobra. Without the Sublime, nature would be stagnant and cloying; without the Beautiful it would be horrendous, too much to bear. This is why a balanced relationship to the natural world—and to life itself, for that matter—requires both rigor and rapture, both war and peace, the relaxant of calm pleasure and the tonic of danger and struggle. If it's all peace, we become effete; if it's all struggle, we become barbaric. There is also Sublimity in Beauty—witness the stallion—and Beauty in Sublimity—witness the tiger.

God too manifests as both Beauty and Sublimity, both Mercy and Justice—which is why an integral vision of virgin nature is the primary support, outside of divinely-revealed religion, for the contemplation of God.

In my lifetime, the 'pastoral' idea of Nature as beautiful and protective has given way to the 'awesome' and 'x-treme' sense of Nature as sublime and terrible, filled with monstrous life-forms and disastrous events; Nature has changed for us from motherly Demeter into wrathful Kali. This reflects our postmodern hatred and fear of sentimentality, our terror of environmental breakdown, and our

almost conscious belief that we really deserve no better. If 'complacency is fear', then the complacency of the 19th and early 20th centuries in relation to nature has now exposed the fear it once
concealed. To lose all sensitivity to the beauty of nature, however, is
to despair of the Mercy of God, and this is a satanic temptation;
may God protect us from falling into it.

⊚

VII

HUMILITY

BUT THE BEAUTY AND SUBLIMITY of the Earth are not there for us to identify with. Those who celebrate nature only to widen the sphere-of-influence of their egos, so they can become Snake Power Man or Magic Eagle Woman, who look to nature as sanctioning them in their vanity and arrogance, will be lucky if they escape with anything less than a serious case of corniness and bad taste.

I'm obviously speaking here mostly of white people who identify with Native American spiritualities, though Native Americans can certainly be arrogant, just as Wasichus can be sincere. I only want to remind the poachers (and they know who they are) that according to the true spiritual dimension of the Native American traditions, to be given one's spirit-name—not to take it on one's own initiative—is to serve one's celestial archetype, not to rip off glamorous energy from the natural world to serve one's ego. In the face of the beauty and sublimity of the Earth, given that we witness them in sincerity, we will know ourselves as small; temporary; nothing much in the total scale of things—and therefore as host to the Great Mystery.

The Latin word for 'earth' or 'soil' is *humus*, from which we get our word 'humility'. When the Lakota go on vision-quest, they do not demand or try to 'access' a vision—they 'lament' for it; they make themselves low, and poor. In the Qur'an, Allah says: 'I am the Rich and ye are the poor.' When Jesus said 'I come to preach the good news to the poor' and 'the poor in spirit shall see God,' it was these 'poor' he was referring to. The *Tao Te Ching* says:

The greatest good is like water.
Water feeds the ten-thousand things, but it does not struggle.

It flows in lowly channels, through places people avoid;
In this it resembles the Tao.
In living, hug the land.
In meditation, plumb the heart.

Accept shame gladly.
Accept suffering as a normal part of life.
What does it mean to 'accept shame gladly'?
Accept being nobody special.

Humbly surrender your life;
Then you will be entrusted with the care of all things.

Jesus, in the Gospels, says the same thing: 'The meek shall inherit the Earth.' The essence of nature, the essence of all manifest existence, is poverty. It possesses nothing of its own. It's like a mirror.

VIII

VIRGIN TIME

WHEN WE RELATE TO THE WORLD of Virgin Nature, our sense of time changes.

The most limited, most alienated form of time is chaotic time, the time of the obsessive-compulsive, or the hyperactive child. Everything in chaotic time is present tense, but there is no continuity, no real 'presence'.

The next larger form of time is linear-historical time. Now there is a past and a future, which allow us to reflect and plan, to imagine alternative possibilities. Still, we cannot really occupy past or future; the only place we can actually live is in the swiftly-passing present, which, if we become too occupied with memory and anticipation, we will forget to occupy.

The time of virgin nature is cyclical time, which is larger still. In cyclical time, past and future are not separated by the present, but form a single field, turning on the point of the present. Since there is nothing, past or future, which will not eventually 'come around', we can become still. Since there is nothing in time that we need to run after, or away from, we can contemplate time instead of struggling with it; we can sit and watch it pass. 'To everything there is a season, and a time to every purpose under heaven.' But it still depends upon us, upon the quality of our attention, whether cyclical time is the threshold of Eternity, or only 'vain repetition'. If we are placed on the edge of the Great Circle, we are in bondage to natural cycles, under the 'regime of Nature', controlled by an external fate (see Chapter X). It was this feeling of bondage, late in the Middle Ages, when the aeon of cyclical time was growing old, that made the aeon of linear-historical time, which began in the Renaissance, feel like a great liberation. So every dissipation feels—at first.

If, however, we are placed at the center of the Great Circle, at 'the still point of the turning world', then some part of our consciousness must be beyond time entirely, since you can only view something whole if you are not identified with it. Beyond this, there is only the Timeless, the Eternal Present, where all is available, and all is empty.

☙

IX

THE NATURE SPIRITS

WALKING IN A FOREST (at least that's where it happens for me), you can learn something about 'planes of being'. What we experience through our five senses is the physical plane. What we experience through dreams and mental images is part of the 'astral' or 'imaginal' plane. And the plane between these two—what we experience in terms of a 'warping' of the physical plane, or as distinct out-there-in-space hallucinations—is the 'etheric' or 'animic' plane. The animic plane is the 'world soul', the 'universal plastic ether', the web of vital, sentient energy that knits together the imaginal plane, where thoughts and feeling are as real as animals, and the physical plane, where a rock is real because you can kick it.

The animic plane is where the nature spirits live, whom the Muslims call the Jinn, and the Celts, the Fairy Folk. They are the demigods, elemental spirits and 'little people' recognized by all cultures except the most materialistic. When one is in a state of alert reverie, or relaxed vigilance, and the face of the forest opens up to reveal the forest *behind* the forest, that is Their world. I mention them not because their appearance says anything particularly significant about the one encountering them, who is neither specially blessed nor specially damned (necessarily), nor because they can in any way act as guides for us, but simply because they are among the sentient races of which the local universe is composed—and because, if we encounter them, and believe what we've seen, we will realize that the world around us is much more alive than we ever suspected.

Their world is none of our business, unless it appears. And, if it appears, it behooves us to practice courtesy: 'Sorry,' we should tell them, 'don't mean to bother you; just passing through.'

In spiritual terms, the world of the elves, the *Sidhe*, the Fairy Folk, the Land of the Ever-Young, is—like the material world—basically neutral. On the other hand, if we are really serious about the spiritual Path, then nothing is neutral. Like the material world, the animic plane is filled with beings who are good, evil, wise, foolish, grave, impish, or simply doing their job. But the fact that they are not material in quite the way we are can make us overly impressed with them; their very strangeness can seduce us. This is why, in terms of a fundamental orientation to the Source and Goal of our being, their world can be a great distraction, even a fatal one.

However, once such an orientation is firmly established, and we are called to begin to discern the Face of God hidden in all things, an encounter with the Folk may—God willing—be part of our response to that call.

For many of us, even those with a basically religious worldview, the natural world seems alienated from God, Who is imagined to be elsewhere. And certainly That One *is* elsewhere than in nature as *materialism* sees it, the natural world as a sophisticated bio-techno-logical device... But to believe in, and then actually encounter, the nature spirits is to destroy the gross materialistic model of nature, root and branch (though a *subtle* materialism, an idolatry of psychic energies, may still remain). So we can say that if one is truly obedient to God's Will, and understands that God in His Essence transcends all that can be experienced through the senses or grasped by the mind, then an encounter with the nature spirits may be a sign of the dawning of the Divine Immanence, of our ability to know God not as elsewhere, but as everywhere.

The danger is, however, that the nature spirits can be immensely fascinating and *diverting*—which means that if we attach to them out of motives of visionary self-indulgence, or hunger for magic power, they will—through no fault of their own—divert us from the Path of God; and there is no greater misfortune than this.

But still... when a spiritual influence comes in from the spiritual Sun, which is the Divine Intellect, and shines down through the angelic hierarchies, down through the astral dimension or imaginal plane, till it reaches the borders of their world, then a perceptual opening to that world can help 'tenderize' us, so that the influence in question, God willing, can come fully into our physical life. So it is best when this opening happens by the Grace of God, without our willful interference—because if we force this opening through harsh artificial means, it may look to them like somebody banging on their door and yelling in a foreign language, late at night. Who can blame them if they draw their weapons? And if the energy-wall between the physical and animic planes is breached, not from above by a descending angelic influence, but from below, by a frivolous curiosity, a well-intentioned but ill-conceived desire to manipulate subtle forces, or a gross hankering for power, then our life-energy will simply bleed away into the branching corridors of the forest, until we become *pixilated*—'pixie-led'. And only an undeserved intervention by angelic protectors can prevent those holes in our field of vital energy from filling with the powers of darkness.

◈

To my knowledge it is only the Tibetan Buddhists, whose tradition is a unique synthesis of a high 'religion of salvation' and a body of ancient shamanic lore, who have been able to fully orient the experience of the nature spirits and their world toward spiritual liberation. When a Tibetan yogi wants to practice *sadhana* in a natural setting, he or she first contacts the nature spirits, informs them of his or her intent, and asks their blessing, just as we might ask a dairy rancher: 'I'd like to sit up on that rock and meditate for a couple of hours—would that be alright?' It's only common courtesy.

◈

X

THE
REGIME OF NATURE

WHEN OUR HUMANITY falls under the power of the ego, Nature is transformed from a manifestation of God into a regime, a system of inexorable physical laws governing the motions of a great World Mechanism. The heartbeat of a vast living being, the Primordial Humanity, is now the pounding of pistons; in Blake's terms, the 'wheels within wheels' of Ezekiel's vision, where all has a common center and that center is everywhere, are transformed into 'wheels without wheels', the interlocking gears of the cosmic machine, where physical events are seen as caused by physical antecedents, not by the Will of God. But since this vision of natural law is too narrow to predict or account for our total experience, it casts *fatalism* as a shadow. What was once known as the creative *shakti* of the Deity is now experienced as a tyrannical Queen, a vengeful and mysterious Goddess of Destiny whose whim is law. What was once Providence is now Fate.

As the Hindus and Buddhists would say, as soon as there is an ego, there is karma; as soon as an identity is posited apart from the Absolute, one apparently capable of independent action, then Fate—the results of actions or intentions too deeply buried in the past (or the unconscious) to ever be untangled—swings into action. And the fact that the clockwork universe of Newton which was dominant in Blake's time has now been replaced by a world of chaos and indeterminacy, the universe of Heisenberg, in no way threatens the regime of Fate, which is all the more fatal now in that its causes cannot be traced even in theory, lost as they are in maddening tininess and daunting immensity. The essence of *materialistic* paganism (not the high paganism of, say, the Orphic tradition) is the worship of this external fate, which is a form of religious *literalism*: If the cycles of the Earth and the astronomical universe are inexorable, then we must serve them in fear. Since they are no longer known as expressions of the Divine within us, orbiting the transcendent Center within the human Heart, but seen rather as emissaries of a mysterious and distant Object which we are commanded to obey but are never allowed to know, then we must sacrifice our children to placate them, or know the wrath of Fate.

And if the Sun is no longer in our hearts, but rules now from the sky above, a jealous and hateful god, then we must tear our hearts out to feed him. Such literalism is not only the disease of a degenerate paganism, however, but of alienated Christianity as well, where inexorable moral law replaces the forgiveness of sins. For fanatical Christians and Jews and Muslims, just as surely as for the pagan Druids, God is a stone on which the human Heart must be sacrificed, so that hearts may become stone in every breast.

But this alienated, literalistic, materialistic regime of an external nature ruled by mysterious Fate is no more real than the human ego, the only source of karma, and its only victim. When God is realized, when the ego is transcended, then karma ends—and the sword Christ came to bring, the only sword sharp enough to cut the chain of *karma* (though other traditions also possess it, under different names) is the forgiveness of sins. When, by the power of God, our sins are forgiven—and when, by the same power operating through us, we find we are able to forgive others—then the world, as we have known it, ends. It was a world of memory, that is, forgetfulness; it was a world of forgetfulness, that is, memory—a universe of grudges and denials. It was a world where both those who remember the past and those who do not are condemned to repeat it, since the first group is ignorant of causes, and so helpless to change them, and the second is obsessed with known causes, and so unable to imagine alternatives. It was a world where the Eternal Present was veiled by space, time, matter and energy.

But now that veil is torn in two. Space is God's form. Time is God's action. Matter is God's substance. Energy is God's power (though God, in Essence, is infinitely beyond that form, that action, that substance, and that power)—and in THIS world, which is the world where you, dear reader, presently stand, or lie, or sit:

Not body, but only body's corpse, is transcended;
Not nature, but only nature's regime, is ended.

XI

FALL AND APOCALYPSE

UP TO A POINT, nature is here for us to use—for food, clothing and shelter. Unfortunately, we never seem to be able to figure out just where that point is. We burn the rainforests. We strip-mine the mountains. But there are other, subtler ways to use up nature, best summed up in the words of an Eskimo *angakok* (shaman), who said: 'The greatest danger to life on Earth comes from the fact that the food of human beings consists entirely of souls.'

What does this mean? Exactly what are these 'edible souls' without which the human race could not survive?

In the 'Garden of Eden', that primordial Golden Age which left few if any historical records, but is found in the legends of almost all peoples, Humanity 'walked with God in the cool of the evening.' We were intimate with the Divine Source of all life; the race as a whole was open to the vision of Eternity, in which all forms and events are known as names and acts of God. Needless to say, in this Terrestrial Paradise all things were seen as living souls—and rightly so. But after we 'ate of the Tree of the Knowledge of Good and Evil', this original unitary vision—the Tree of Life—was covered over, and we fell into a limited, essentially dualistic mode of perception. We still believed in God, and often saw Him, But our original unbroken intimacy with Him was lost. At this point the knowledge of good and evil, which in Eden was useless to us, suddenly became necessary: Since God was perceived as being at a distance, we now had to choose those things which brought us closer to Him, and reject those things which pulled us further away. Nonetheless, we still saw all things as being, or having, souls—but since we were no longer in unbroken, conscious contact with the Divine Source of all life, we began to turn for our psychic and spiritual life not to their real Source, but to the secondary reflections of that Source in the forms of the world around us.

And we've been doing it ever since. In an earlier age (perhaps the 'Silver' one, of which present-day shamanism would be a distant survival), we consciously recognized that we were feeding upon souls, and so consciously compensated for it through sacrifice and purification, especially after taking animal or human life. We said, in effect: 'Almighty God, we know that You are the only source of life; therefore forgive us if we can't always see this; forgive us if we must now perceive the basis of our life in this world as something other than You. Save us from the spiritual and physical dangers of this way of seeing things; don't let us take You for granted; don't let us forget You.' But in later ages, and most especially in this present Age of Iron, we progressively forgot the Divine Source of all life; we fell more and more deeply into materialism, and into the cold-heartedness and ecological destruction that are materialism's fruits.

To see an object outside yourself, rather than a Reality that both contains you and is contained within you, as the ultimate source of your life—whether it be a buffalo, a gold mine in the Black Hills, a high-paying job, your mother, or all of material nature—is to start eating souls. It is done through identification. I add something 'out there' to my psyche in order to increase its vitality and widen its field: a totem animal, a familiar spirit, the heart of a slain enemy—a Jaguar, a lover, a first-growth forest, a rival corporation, a neighboring country—all of them I see as reservoirs of desirable soul-substance that I wish to consume.

But identification is not without consequences; it has to be paid for. We pay for it by losing a part of our soul for every part of the outer world we incorporate. In the words of Jesus, 'What profit it a man if he gain the whole world, and lose his soul?' To 'eat of the Tree of the Knowledge of Good and Evil' is to fall from unity-of-perception into duality. And the first step on this descending road is the subject/object split, the distinction between seer and seen.

The subject/object split is another name for the illusion of EGO; the belief that I am self-subsistent generates the perception of the world as other than, and opposed to, the self. The ego is generated through identification, and identification has two parts to it—what psychologists call 'projection' and 'introjection'. To 'project' is to see something that is really in me as being in an outer thing, person or situation; to 'introject' is to see something that is really in an outer thing, person or situation as being in me.

In the Primordial Unity, the experiencer, the thing experienced, and the experience itself are One—which means that there is no 'unconscious', no storing of impressions, no memory. But when the experiencer and the thing experienced are separated, then experience becomes incomplete. And just as incomplete combustion produces smoke, so incomplete experience produces memory, which is why our direct experience of God, Who is in reality the only complete experience, is never remembered. (The generation of memory is the same thing as the perception of linear time. Complete experience is global; to the degree that experience is incomplete, it must become sequential.)

☽

Incomplete experiences, stored as memories, are 'repressed', they drop out of consciousness—and as Carl Jung said, whatever is repressed is projected. Because we store impressions in unconscious memory, we project them on the outer world; consequently we edit our present experience according to the 'karma' of our past experience, thus generating more karma for the future. And whatever object we project upon, we also introject: If I project my own potential or forgotten strength on a friend of celebrity, that person is inside me—which is why to flatter someone is to devour them. If I project my anger on the world, then I see and am filled with an angry world (which is not to say there is no anger in the world, only that another name for projection is 'unconscious selection of experience'); and if I project the lost Unity of subject and object on the screen of material nature, then I will attempt to incorporate that nature, to devour it whole, in order to restore that Unity. This, of course, does not work. The more we devour the outer world, the more we lose ourselves in it, because not only does every projection lead to an introjection, every introjection also leads to a further projection. And so the world also devours us. And the more of ourselves we lose in the outer world, the bigger the hole we leave in our souls—a hole that needs to be filled, because 'nature abhors a vacuum'. So the next time around we are even hungrier, consequently we devour an even bigger chunk of the world, leaving that world even hungrier for us that it was before. Our hunger always increases, because all we are eating is hunger.

The only way out of this truly vicious circle is to remember God as present in this moment. The practice of this 'remembering' (which is the exact opposite of the storing of memories) is the Spiritual Path. When we remember God, the process of identification ends. When identification ends, the process of projection-and-introjection is reversed. Simultaneously we realize that the 'us' in question, our 'self' as separate from our 'world', is really nothing in itself; it is merely a mass of stored experiences. Realizing this, we release these stored experiences, these 'introjections', back into the 'world'—which, in consequence, is no longer a 'world' separate from our 'self', but a seamless field of experience which flows in upon us as an ocean of vital, sentient energy (what the Hindu Tantrics call *shakti*) in which experience, experienced and experiencer are one: the Primordial Unity restored.

In the Book of Revelations, the release of stored memories is symbolized by 'the resurrection of the dead'. The unconscious, the place where memories are stored, is the 'sea' which shall 'give up her dead' (Rev. 20:13), after which there shall be 'no more sea' (Rev. 21:1). And the dawning of *shakti* is symbolized by the descent of the Heavenly Jerusalem (Rev. 21), the 'bride of the Lamb' who is the 'new heaven' and the 'new earth' around us, perfectly united with the Divine Consciousness (the Lamb) within us.

The only way to heal the environment is to stop devouring the world. The only way to stop devouring the world is to remember God as present in this moment.

᭥

XII

TECHNOLOGY
AND SORCERY:
THE POWER TRAP

PEOPLE USUALLY THINK OF technology and sorcery, engineering and nature-magic as poles apart. One is gross and destructive; the other is subtle, fascinating and pulsing with life. However, this antithesis can only be taken so far. Magic, we should remember, was the historical ancestor of technology; a certain similarity of outlook still remains—a similarity which is actually increasing, now that the technological manipulation of brain-function has become one of the scientific 'black arts'.

Some shamans, it is true, are faithful servants of God and their people—but there is a certain class of sorcerers, sometimes known as 'wizards', who burn and strip-mine the subtle energies of the natural world in the name of personal power. They enslave the spirits of the Earth instead of befriending them; they rip off the etheric plane just as engineers and technicians rip off the physical plane, and with much the same effect: for the Earth, damage and depletion; for themselves and their victims, hearts turned to stone.

When the subtle forces of the natural world are no longer seen as symbols and emissaries of the Great Spirit, but rather as resources to be exploited by the wizard-ego, then the basic set of nature-magic has become indistinguishable from that of gross technology. To worship God is to be allowed to use nature; to attempt to use nature without worshipping God is to destroy it, and be destroyed ourselves.

Merlin, for good or ill, was a wizard, albeit a 'white' one; he used nature for his own ends. He created the fellowship of the Round Table, and that was a worthy creation. But to the degree that he used the natural world, he was also used by her; to the degree that he enslaved the elemental powers of nature, he was also enslaved by them. The story of his end is exquisitely accurate: In his old age he fell in love—idiotically, hopelessly, as old men sometimes do—with a young girl: Niniane.

She wrapped him around her little finger, till he had taught her all his magic; and because Merlin retained a part of his wisdom he knew exactly what was happening to him, but was powerless to prevent it. Finally Niniane, using one of Merlin's own spells, cast him into a magic sleep, or imprisoned him in a magic tower, or entangled him in a magic whitethorn hedge, from which no-one could ever free him—not even herself. He who was once a man was now nothing but a voice; locked in a powerful magic glamour, a superhuman trance with subhuman consequences, he awaits the Day of Judgement.

◈

XIII

POETIC IMMORTALITY: THE ESTHETIC TRAP

THERE IS A CERTAIN WAY of relating to Nature, usually confined to that class of people known as 'romantics'—a way of receiving the blessing of the *regime* of Nature—that turns those so blessed into gods and goddesses, or deified heroes, or holy ancestors. All throughout the ancient world there were shrines dedicated to heroes or heroines who had become gods and goddesses after their death. And undoubtedly these great men and women were in some sense the 'children' of, the emanations of, those deities, or archetypes, or Names of God which they had served during their lives. Of course it is true that no-one who has not encountered and willingly placed himself under the power of his or her archetype, or genius, or tutelary deity can lead a fully human life. But there is also a way in which one's relationship to his or her tutelary deity, especially if it is unconscious, can diminish that humanity. In other words, when we dream of being demigods because we are afraid to live as human beings, then, sometimes, the Regime of Nature will grant us that wish. During life we will become 'fey', a word meaning 'fated', or 'dedicated to a particular god', or 'marked out for sacrifice to a particular deity', and which is also related to the word 'fay', meaning 'fairy'. (Thus Morgan LeFay, for example, was both 'Morrigan the Fairy' and 'Morrigan the Fate'.)

The one who was *fey* had one foot in the other world, and was considered doomed to die within a stated period. And one of the most common ways of becoming fey like this is through the worship of the aesthetic beauty and fascination of Nature as something separate from our full humanity—which is why this condition was traditionally, and often still is, a particular occupational hazard of poets. Those dedicated to the Beauty and Sublimity of nature as values in themselves, rather than as manifestations of nature's Divine Source, in light of our full humanity, attain 'poetic immortality'. In life, they are *fey*. In death, they become prisoners of the Nature Spirits; dwellers in the 'Land of the Ever-Young'; hostages of the *Sidhe*.

Thankfully, this isn't Hell. Regretfully, this isn't Purgatory either. In Purgatory more is going on, there is more change, more development, more purifying suffering. The 'Land of the Ever-Young' is, rather, the 'Limbo' of Dante's *Inferno*, where all the classical poets live, surrounded by the beautiful forms of nature, but cut off from the light of the Holy Spirit. Nothing much changes there, millennium after millennium. In the Irish rendering of it, the poets and heroes feast on 'swineflesh, milk, and mead', and make love, forever, in endless glamour, subhuman and superhuman at the same time, to the beautiful women of Faerie, accompanied by the singing of exquisite poems, set to a music of unearthly loveliness, and heart-rending nostalgia. But if anyone returns from that 'all too divine' land to the world of earthly reality, if poet or hero ever sets foot, like Oisin did, on that all too human ground, he will shrivel up into a pitiful old man, and die in minutes. It had seemed to him as if only months had passed under that Faerie hill; in reality, centuries had passed. He had allowed himself to be beguiled by an inhuman beauty, only to learn that the sole way to the Transpersonal is through the Personal, that the only path beyond the human leads straight through the human—which is what Jesus meant when he said: 'None come to the Father but through me.' Aesthetic beauty is merely outside of time; love, however, even though it lives in the heart of time, is in Eternity already.

ⓢ

The Vedas speak of two paths to be taken by the soul after death: the path of the Waning Moon which leads to the Way of the Ancestors and ultimate rebirth—*pitri-yana*—and the path of the Waxing Moon, which leads to the Way of the Gods, the Door of the Sun, and final liberation—*deva-yana*. Poetic immortality is allied to the Way of the Ancestors. Just as the ancestral ghosts are kept in more-or-less human form by the offerings of their living descendants, by their active memory, so poetic immortality lasts only so long as the words of the poet continue to be recited or the deeds of the hero celebrated. To attain this immortality is to enter the human memory, where the remembered dead are stored up like seed-corn. They are the souls of children, waiting to be born—and just as the memory of a great-grandmother is redeemed from the collective tomb of the ancestors by being born as part of the psychic 'sheath' of her own great-granddaughter, so a hero is reborn in whoever his remembered deeds have moved to new exploits, and a poet in whoever his remembered words have inspired to fresh creation. As it says in the Welsh poem 'The Battle of the Trees':

I was a word in a book;
I was a book originally.

But when will the immortal poet, crowned with laurel, break out of that charmed circle and return in *human* form? Is it not better to be a real man or woman, no matter how burdened, than a voice upon the wind?

◈

XIV

THE
TERRESTRIAL
PARADISE

BUT ON A HIGHER OCTAVE, the Land of the Ever-Young is not simply an island of irrelevant lyrical beauty, heavy with nostalgia, moving off at an oblique angle from the current of Divine manifestation which the Bible calls the Tree of Life, but the Terrestrial Paradise itself, the Imaginal Earth, where the Tree of Life bursts into subtle material form. The Terrestrial Paradise is indeed the Earth—but it is a vastly greater Earth than the five material senses report. It is what the Muslims call the 'Earth of Hurqalya' or the 'Eighth Clime', as if it were a mysterious eighth continent that appears on no map. It stretches into dimensions of immensity that are effectively closed to us, most of the time; yet those dimensions keep established outposts as close as the nearest 'tenanted grove', the nearest piece of vital, sentient Nature, where, from behind the transparent curtains of the forest, the Nature Spirits peer out at you & me peering in.

⑨

In traditional Catholic theology it is said that when humanity fell, disorder was introduced into nature. What does this mean? On the most comprehensive level, it means that when human consciousness falls under the power of ego, the natural world is reduced from a manifestation of God to an object of the fears and desires, the disgust and longing of that ego. But in a more limited sense, it means that some of the Nature Spirits are fallen, and some are not. In other words, it is possible for the Fairy Folk, just as it is for men and women like you and me, to forget their Creator, and so begin to believe that they are self-created. This is what the Muslims mean when they say that some of the Jinn are Muslim and some are not. And those who are not, those who worship their own kings and queens but have forgotten their Divine Source, the Fountain and the Tree of Life, are the ones who create and people that hell reserved for disobedient poets, that fallen paradise of aesthetic beauty and spiritual despair.

But on another level, we can say that those of us who believe we are self-created—those immersed, that is, in egotism—will, if by some fluke or powerful artificial means we are able to access the world of the Nature Spirits, encounter only those spirits who believe themselves to be self-created too—this being one of the many roads to the Kingdom of Darkness. On the other hand—on the opposite horizon of things, the *Eastern* one—those of us who know ourselves as created instant-by-instant by the Divine Source of all life, the breath of God clouding the mirror, will encounter only those spirits who also know this, those who worship Allah, the ones who cluster like vivid wildflowers, prismatic dewdrops, busy bugs and splinters of clear quartz crystal, around the Roots of the Tree of Life. (Even these, however, are dangerous to encounter, since their energy tends to dissipate the human psyche and distract us from fulfilling the human trust. There was once a Sufi saint who, as he performed his daily prayers, saw that some of the Jinn were praying with him. These were obviously the pious, Muslim Jinn—yet he asked them kindly to go and pray somewhere else, since it was both his intent and his duty to concentrate upon God Alone.)

If you spend most of your waking life trying to establish and defend your own identity, you will see Nature as a mass of mechanistic material processes, as something to be engineered. If, however, you spend your time contemplating the living Source of your being, and acting in spontaneous obedience to that Source, then you will see all Nature as doing the same. And you will be right.

⑨

When Adam ate the fruit, whole sections of Nature were cut off from the Tree of Life and left to drift, hopelessly, into space, time, matter and energy—and yet, not all of Nature fell. One of the clear-est proofs of this is the paradise of the Celtic Christian monks, who, in their radiant simplicity, sat stitching the realms of the *Sidhe*, the Fairy Folk, back into the living fabric of Eternity. In his heroic inno-cence, one of them wrote:

I wish, O Son of the Living God, ancient and eternal King
for a secret hut in the wilderness that it may be my dwelling.

A very blue shallow well to be beside it, a clear pool for
the nurture of many-voiced birds, to shelter and hide it.

Facing south for warmth, a little stream across its enclosure,
a choice ground with abundant bounties which would
be good for every plant.

A few sage disciples, I will tell their number, humble and
   obedient…
Six couples in addition to myself, praying through the
   long ages [sic!]
To the King who rules the sun.

❧

XI

WHO IS THE EARTH?

To come face to face with the Earth not as a conglomeration of physical facts but in the person of its Angel is an essentially psychic event which can 'take place' neither in the world of impersonal abstract concepts nor on the plane of mere sensory data... the perception of the Earth Angel will come about in an intermediate universe... a universe of archetype Images, experienced as so many personal presences. In recapturing the intentions on which the constitution of this universe depend, in which the Earth is represented, mediated, and encountered in the person of its Angel, we discover that it is much less a matter of answering questions concerning essences ('what is it?') than questions concerning persons ('who is it?' or 'to whom does it correspond?'), for example, who is the Earth? who are the waters, the plants, the mountains? or, to whom do they correspond? The answer to these questions causes an Image to appear and this Image invariably corresponds to the presence of a certain state. This is why we have to recapture here the phenomenon of the Earth as an angelophany or mental apparition of its Angel in the fundamental angelology of Mazdaism as a whole, in that which gives its cosmology and its physics a structure such that they include an answer to the question, 'who?'

—Henry Corbin, *Spiritual Body and Celestial Earth: From Mazdean Iran to Shi'ite Iran*

THIS MATERIAL EARTH is not the Great Mother, the Great Goddess. She is limited in space and time; she has boundaries; she has a beginning and an end. In this she is, in a certain sense, an individual

like any one of us. The Great Goddess is the feminine face of the
Absolute, not a limited entity like a planet or a supergalaxy. She is
the Reality underlying all manifestation, the infinite Matrix of all
becoming, the 'Void eternally generative'. She is that aspect of the
Godhead which, because it is completely beyond form, can give
birth to all forms. She is the Divine as Substance—infinite, infinitely
subtle and infinitely pervasive—of which the entire space-time
matrix, the universal field of matter/energy, is only a partial rendition, a later, and lesser, echo. That's how great She is—so great that
the Earth, and even the total material universe, are like grains of
dust in Her vastness. This Divine Substance is what the Lakota call
'Grandmother Earth'. This material Earth, 'Mother Earth', is her
daughter.

But there is more to this limited, individual Earth than our five
senses report, just as there is more to a human being than his or her
physical body. This fragile, mortal Earth has, and in one sense is, an
immortal soul—which is why, according to Mazdean (Zoroastrian)
belief, the Earth is an Archangel. Her name, in the Zend Avesta, is
given as *Spenta Armaiti*, 'perfect thought'. Humanity is her son; and
her daughter is *Daena*, '(perfect) action'. Daena is, in one sense, the
sister of Man; in another sense, she is his daughter. Man is born of
the Earth as Spenta Armaiti; his realization of her as his archangelic
mother becomes his encounter with Daena, his own human soul;
and his soul, as 'action', is also his product or manifestation, his
'daughter'. Spenta Armaiti/Daena is thus the *shakti* of humanity,
what William Blake called our 'emanation' and personified as Jerusalem, both City of God and City of Man, at once the product of our
creative labor, and our eternal Bride. As Spenta Armaiti she is the
total unborn or pre-eternal potential of what it would be to be
humanity incarnate on Earth; as Daena, she is the total realization
of this potential through divinely-inspired human labor.

And so the Earth, in one sense, is the set of all the potential experiences of all the conscious beings who are her children, of which
Humanity is the center and synthesis: She is the *shakti* of Man, as
Grandmother Earth, her mother, is the *shakti* of God. She is all we
could ever be, in the perfect realization of our human integrity,

there before us already—arisen, from the depth of our soul, like Dante's Beatrice, as the very Angel and Image of what a human soul really is. This material Earth is like a distant memory of her—and as we *remember* the Earth, as we make her present to ourselves by ourselves becoming present to God, She opens, like a door. Within her is Earth, the living conscious being. And within that Earth is Earth, the eternal archangel. And within that Earth is Earth, the Grandmother, the manifestation of the Divine Infinity, whom the Muslims call 'The Guarded Tablet' or 'The Mother of the Book' and the Jews 'The Torah'—those 'waters' upon which the Spirit of God moved in Genesis. There are higher Earths, each one inhabited by a higher Humanity. As those Earths are alive within this Earth, so those Humanities are alive within us.

◈

XVI

CHRISTIANITY: UNWORDLY, NOT UNEARTHLY

In traditional Christian theology, we hear a lot about 'the world' as a temptation, a source of sin. It's pictured as something we should avoid, something dangerous to touch. Contemporary westerners, living amid the ruins of a Christian culture but with very little understanding of Christian tradition (even among many Christians), project their feelings of alienation from nature on Christianity. They think 'the world' means 'the Earth'.

They are wrong. 'The world' is not the planet but the established system of things, the conventional collective mindset and the moral and socio-economic establishment that's based on it. We don't call people who are in touch with nature 'worldly'; we call them 'earthy'. Traditional Christian monks, working their fields with simple tools and living in bare stone cells, were and are earthy, not worldly.

In Catholic doctrine, the World is one of the three sources of temptation, the other two being the Flesh and the Devil. The Flesh is a fragment of who we are as human beings; it's the mass of our desires for pleasure and security, cut off from our human wholeness and operating on automatic pilot. (There is nothing wrong, of course, with pleasure and security—in moderation—but we all know what happens when our struggle for these goods becomes obsessive, takes over our whole lives, and blots out everything else.) The Devil is that transpersonal spiritual power who exploits the possibility of our denial of our God-given potential for human wholeness on every level, including that of the misuse of spiritual gifts.

The World is another fragment of us. It's our competence and sense of responsibility operating apart from our wholeness, and at odds with it. It is, in other words, the addiction to power, and it works in collusion with those collective social beliefs that determine whether or not we are 'the right kind of person'. The World tempts us to be who we are supposed to be in the eyes of society, not who we really are in the eyes of God. It is materialism, worldliness, the way of the world. It is not the Earth. It is what destroys the Earth.

☙

The first chapter of Genesis, verses 27 and 28, is as follows:

> So God created man in his own image, in the image of God created he him; male and female created he them. And God blessed them, and God said unto them, Be fruitful, and multiply, and replenish the earth, and subdue it: and have dominion over the fish of the sea, and over the fowl of the air, and over every living thing that moveth on the earth.

Those who hold the Judeo-Christian tradition responsible for our environmental destruction quote this passage more often than any other. In particular, they don't like the idea that God has given us 'dominion' over the animals, and commanded us to 'subdue' the Earth. They believe that earlier traditions, possibly Goddess traditions, were more respectful of living things. But in reality, the image of the 'subduer of beasts', the Animal Master (cf. Joseph Campbell, *Primitive Mythology*), whether god, goddess, yogi, or shaman, is found everywhere: in the Neolithic, in the Paleolithic, in the Indus culture, the Mesopotamian culture, the Celtic culture, the Native American culture—it's nearly universal. The nuance of the Genesis account may be different, say, than that of the Celtic account, since the nuance in every account is different, but the myth is, in essence, identical. Without 'dominion over the beasts'—without (in the Babylonian rendition) Marduk's slaying of Tiamat—there is no meat, no leather, no dairy products, no fish, no poultry, no herding cultures like the Masai, no hunting cultures like the Inuit; the earliest magic of which we have any knowledge was (if we read the walls of the Paleolithic caves correctly) precisely for the purpose of gaining 'dominion over the beasts'. And without a 'subduing of the earth' in one form or another, there are no cultivated vegetables, no grains, no pottery, no brick or adobe or stone or wooden houses— no human life as we know it.

And note: God, in Genesis, commands us not only to subdue the Earth, but *replenish* it. In the same way, Stone Age hunting magic usually includes some form of dedication or release of the soul of the slain beast to replenish the herd. The only function of farmers and stockbreeders (however well or poorly they fulfill it) is to replenish the animal and vegetable kingdoms; that's what agriculture and animal husbandry are.

This means that the culture of the earlier Goddess-religions, like every other culture we know or can conceive of, was based on subduing the Earth, and replenishing it, primarily through agriculture. Some anthropologists even conjecture that women invented agriculture, that farming evolved from primitive gathering, done mostly by women, just as herding developed from hunting, done mostly by men.

Now to 'subdue' the Earth in any real sense clearly requires a balanced relationship to natural resources and some form of population control. If overpopulation leads to massive famine and pollution, if the icecaps melt due to global warming, then we have obviously not subdued the Earth. Quite the contrary: we have driven her wild. Nor is Genesis ignorant of the dangers of such imbalance, especially in terms of urbanization. In the story of Cain and Abel, God accepts the offering of Abel, the herder, while rejecting that of Cain, the farmer. And it is Cain who, after committing the first murder, goes on to found the first city.

Furthermore, like everything in the Bible, Genesis 1:27–28 can be read on an inner esoteric level as well as an outer exoteric one. Esoterically speaking, the 'earth' to be subdued and replenished is the human body—not as a material object, but as a field of psychic experience. The 'fish', 'fowl', and 'every thing that moveth upon the earth' are the psycho-physical powers which manifest within the body, and give life to it. The *fish* are our unconscious, instinctive energies; the *fowl* are our higher thinking functions; the living things which move *upon the earth*—the most conspicuous of which are the mammals— represent our warm, emotional, 'mammalian' nature. These are the 'living things' which humanity, on the inner level, is commanded to replenish and subdue; it is the threefold challenge of contemplative asceticism to overcome the tyranny of instinct, thought and emotion without depleting our vital energies, destroying our ability to think, or petrifying our affections. This alchemical 'great work' is possible because our integral Humanity is on a higher level than the 'natural' soul in us which feels, thinks and instinctively reacts; it is the function of this higher indwelling spirit to 'watch over' our natural inclinations, to cultivate, refine and balance them, and so accomplish what the Plains Indians call 'balancing one's shield', the shield being the four-armed mandala of our psycho-physical faculties, and the life situations that correspond to them.

In Romans 1:20, St. Paul says:

Ever since the world began, [God's] invisible attributes, that is to say his everlasting power and deity, have been visible to the eye of reason in the things he has made.

This passage alone suffices to demonstrate that Christianity is not somehow the enemy of nature, since it knows the natural world as a field for the manifestation of the 'everlasting power and deity' of God. This understanding of nature as sacred to God can be found in the Church Fathers as well. Maximos the Confessor (AD 580–662) says:

The world is one ... for the spiritual world in its totality is manifested in the totality of the perceptible world, mystically expressed in symbolic pictures for those who have eyes to see. And the perceptible world in its entirety is secretly fathomable by the spiritual world in its entirety. ... The former is embodied in the latter through the realities; the latter in the former through the symbols. The operation of the two is one.

Isaac of Nineveh (7th century) says:

> What is purity, briefly? It is a heart full of compassion for all of
> created nature.... And what is a compassionate heart? It is a
> heart that burns for all creation, for the birds, for the beasts....So
> violent is his compassion ... that his heart breaks when he sees
> the suffering of the humblest creature.

And Dionysius the Areopagite (c. AD 500) says:

> It is ... false to repeat ... that it is in matter as such that evil
> resides. For to speak truly, matter also participates in the order,
> the beauty, the form.... How, if it were not so, could Good be
> produced from something evil? How could that thing be evil that
> is impregnated with good?

So Judaism and Christianity are not essentially anti-nature, even
though Western Christianity was ultimately unable to withstand the
secularizing tendencies which, beginning with the Renaissance, led
to the attempt to 'conquer' nature. And the fact that those searching
for a sacred view of nature in the post-Christian west have felt it
necessary to seek the vision of Maximos among the Native Americans, that of Isaac among the Buddhists, and that of Dionysius
among the neo-Pagans, shows just how post-Christian the West
actually is, and how little the western Christian remnant understands its own doctrines and traditions.

The Judeo-Christian-Islamic tradition, the tradition of the 'Abrahamic' religions, is not the cause of our environmental crisis. The
materialistic side of Greek philosophy, which began to unduly influence the West when scholastic Aristotelianism won out over Neo-
Platonism, and triumphed when, in post-Renaissance Europe, it
gave birth to experimental science, is. The Catholic Church tried to
block this development. She failed, not because she was essentially
anti-nature, but because she had lost the dimension of sacred science that could oppose the growth of materialistic, secular science.
Perhaps, given that she no longer had room for the complete
expression of her own tradition, this failure was inevitable—but to
blame her for the triumph of what she fought tooth-and-nail to
prevent, is historically ignorant, to say the very least.

Science and technology are with us; they won't go away. It we can regain a spiritual vision of nature as a direct manifestation of the invisible Source of all life, then they will serve us. If not, they will destroy us.

❧

The following passage is from Revelations 11:18. The 'dead' referred to are not only those who have passed on, but also those whose souls are dead in this life, whose spiritual eyes are blinded, causing them to destroy all they see:

> .... thy wrath is come, and the time of the dead, that they should be judged, and that thou shouldst give reward unto thy servants the prophets, and to the saints, and them that fear thy name, small and great; and shouldst destroy them which destroy the earth.

❧

XVII

NATURE, ART, AND ALCHEMY

THE IMAGE OF HUMANITY as the destroyer of nature is so vivid in our time that we have forgotten the traditional doctrine that man is the perfecter, and also the perfection, of the natural order. An acquaintance of mine, an equestrienne and scholar of horsemanship, once said to me:

> Whoever sees a finely-bred and trained Spanish horse next to a wild mustang, no matter how ignorant of horsebreeding he or she is, will recognize at a glance that the blood horse is the more beautiful. Horsemanship is a spiritual art, because to train a horse correctly, so that animal and rider respond as one, is also to train one's lower soul—not by grossly dominating it, but by teaching it to follow, willingly, the human spirit that rules it.

This is alchemy: Man is both the substance to be refined and the crucible in which the refinement takes place; God is the Refiner, and the Fire. The spiritual secret of any traditional craft is: As nature is shaped and refined, so is human nature. As every traditional craft is an art, so every art is a craft; it is an alchemy which works to synthesize beauty and use, as they are synthesized in Virgin Nature, and in so doing to purge the soul of the grossness or capriciousness that could conceive of an ugly usefulness, or a useless beauty. And the subtlest of the traditional crafts is contemplation, by virtue of which, through the purification of human consciousness, the natural world attains its highest development as God's Mirror, God's Book.

A key concept in alchemy is the Aristotelian polarity between *matter* and *form*. Form is the spirit or essence of something; matter (in the traditional, not the modern sense), is the formless, receptive field where that form manifests. The idea is to unite the two, so that matter is spiritualized and spirit embodied. Related to this polarity, but on a higher level, is the polarity between *essence* and *being*, as expressed by Aristotle, which is a central idea in a great deal of Islamic philosophy as well. If there is an alchemy of form and matter, there is also an alchemy of essence and being; the alchemy of form and matter is in fact possible only because matter, since it is as far below form as being is above it, can act as a representation of being on a lower level.

Being is the 'isness' of things; essence is the 'whatness' of things. That a rock *is* its being; that a rock is a rock, and nothing else, is its essence. Now this way of thinking will sound to many, as it once did to me, like a meaningless mind-game. After all, since no thing exists that has being without essence (since it wouldn't be anything) or essence without being (since it wouldn't be), and given that being and essence can only be separated within the mind, then why separate them? Why, except to create a barren illusion? The answer to this question, or one answer, is—as Blake would say—'to cleanse the doors of perception.'

If you drive by a redwood tree that you've whizzed past a thousand times before, you will tend not to really see it; you will take it for granted. However, if one day you were to see a redwood tree, normal in every way except that it is hanging in mid-air, it would rivet your attention. You would see it in dazzling clarity and incandescent detail. The 'real' tree was of no particular interest to you, but this 'unreal' tree, this *apparitional* one—how vivid it is; how *real*.

The Beat Generation poet Lew Welch wrote a poem called 'Wobbly Rock' about a large rock on the California coast that moves when hit by waves; he used to sit on it to meditate. In the poem he poses the following riddle:

Dychymig Dychymig: (riddle me a riddle)

> Waves and the sea. If you
> take away the sea

Tell me what it is

It took me fourteen years to solve this riddle—or only a moment.
I'll give you the answer a little further on... but elsewhere in the
poem, Lew says:

> Sitting here you look below to other rocks
> Precisely placed as rocks of Ryoanji:
> Foam like swept stones.
>
> > (The mind getting it all confused again:
> > 'snow like frosting on a cake'
> > 'rose so beautiful it don't look real')
>
> Isn't there a clear example here—
> Stone garden shown to me by
> Berkeley painter I never met
> A thousand books and somebody else's boatride ROCKS
>
> > (garden)
>
> EYE
>
> > (nearly empty despite this clutter-image all
> > the opposites canceling out a
> > CIRCULAR process: *Frosting-snow*)

What he's saying, literally, is that the rocks on the California coast,
with white foam between them, remind him of the rocks at the
famous Zen garden of Ryoanji, placed on a field of raked white

gravel—a garden, which, furthermore, he knows only through a painting of it—but then (or simultaneously) he remembers that Ryoanji itself was made that way to remind us of the seacoast. Looking at the California coast, he sees (simultaneously) in his mind's eye the rock garden made to suggest the Japanese coast of the same ocean, a garden constructed in the Taoist manner, 'deliberately unintended' (as he says in another poem)—and so the rocks and white foam light up in his vision, in a moment of what Joyce called 'aesthetic arrest'; they have become more apparitional, more real. The artificial, imagined garden that is and is not the seacoast (where, as in all gardens, art and nature unite) unites with the 'real' seacoast—and the effect of this on the poet's perception is that the ROCKS and his EYE unite. Subject and object are transcended, or become one; the separating ego is dissolved. This is one of the ways in which art serves contemplation.

Furthermore, to say 'snow like frosting on a cake' is both to show again how an aesthetic image can 'cleanse the doors of perception' by simultaneously separating from and uniting with its object, and to demonstrate how a vulgar, cliche'-ridden perception (educated by the *Reader's Digest*) will always miss this, while 'rose so beautiful it don't look real' both expresses the 'metaphysical transparency of phenomena' seen as they really are, and shows how a vulgarized perception—the kind that likes artificial flowers better than real ones—will miss this too.

Elsewhere, Welch defines what he is trying to achieve in these terms: 'I try to write accurately from the poise of mind which lets us see that things are exactly what they seem.'

This is an extremely subtle concept. It seems like naive realism, but it is really the furthest thing from it—except that the most exalted perception, or spiritual state, is finally indistinguishable from the most common. Ibn al-'Arabi's Sufi 'people of blame' look the same as ordinary, simple believers; Kierkegaard's 'Knight of Faith' is like a normal, happy-go-lucky carpenter or fisherman; the most advanced Zen sage, in the last of the 'Ten Ox-herding Pictures', mingles with the crowd in the marketplace and in no way stands out. Likewise in the world of perception: If things are only what they are, heavy literal lumps (as seen by the naive realists, the simple believers in what they see), then they can be taken for granted, and so remain hidden. They are in *sangsara*. If things are 'literally' illusions (as seen by the spiritual travelers, whose vision penetrates to the Real hidden behind phenomena), if they are mere seemings, then we can safely ignore then, and concentrate on the Real, on *Nirvana*. But if they really are what they seem, then they have all the radiant transparency of apparitions, plus all the solidity and presence of real things. Redeemed from both heavy literalism and subjective fantasy, they act as mirrors reflecting the Divine Nature: *Sangsara is Nirvana*.

The Buddhists say that the intrinsic 'emptiness' of things (*shunyata*) is not other than their essential 'suchness' (*tathata*). Shunyata is being, since pure being is empty of specific determinations, empty of essence, While *tathata*, the property of things by virtue of which they are exactly what they are, is essence.

Now, back to the riddle:

> Waves and the sea. If you
> take away the sea
>
>
> Tell me what it is

The answer is: If you take away 'the sea' from 'Waves and the sea', you get 'Waves and', which, to the ear, is also 'waves sand'—so the solution is something anyone who has seen a sandy ocean beach has seen: the pattern of waves, or ripples, left by the ebbing tide in the drying sand. This stationary wave-pattern is *tathata*, essence. The absent sea is *shunyata*, being—or void. To separate essence from being like this, and reunite them on a higher level, is what all true art does, to free us from our habitual ways of looking at things and cleanse the doors of perception.

The Chinese landscape painter, say of the Sung period, renders his pine branches, waterfalls and misty crags simply by removing the *being* of his subject, and leaving only the *essence* (though not, of course, the being of the painting itself; the essence of his subject, first given being by water, timber, rock and air, is now reflected in ink and rice paper). The great classical Chinese or Japanese painter does not try to *reproduce* nature, like the 'realistic' or 'naturalistic' artist, but rather makes a painting which, because it is obviously an 'apparition', an image, and not an imitation or counterfeit of a real thing, thereby reveals the essence of its subject—so that, when we find ourselves walking through a landscape of pine trees and water-falls and misty crags, and suddenly recall such an image, it immedi-ately superimposes itself upon and blends with the picture painted by our senses, since there is no 'rivalry of two beings' to prevent it. We suddenly witness a world in which essence fully reveals being, a world where 'things are exactly what they *seem*.'

This, again, is alchemy. Being and essence, in their fallen state, where manifestation obscures rather than reveals the Principle manifested, are like Mercury and Sulfur chaotically mixed or crushed together, producing that state which the alchemists call 'lead'. For this 'lead' to be transmuted into 'gold', where manifesta-tion reveals the Principle rather than obscuring it, Mercury and Sulfur—matter and form—being and essence—must be separated, clearly distinguished, and then reunited on a higher level.

This is the alchemy which all true art serves. As the human sub-stance is purified and refined through spiritual practice, the vision of nature becomes a theophany, a revelation of God. As the vision of nature is purified through art, the human substance (Salt) is refined as well, so that it can more perfectly reflect its Divine Source. To remember God in the invisible, transcendent world is to gain the ability to see Him in nature. And when we see God in nature, then the natural world—like the wild beasts who come to lie at the feet of a realized saint—sees God in us. As the Emerald Tablet says:

> In truth, certainly and without doubt, whatever is below is like that which is above, and whatever is above is like that which is below, to accomplish the miracles of the One Thing.

֍

XVIII

CONTEMPLATING NATURE
AS
JÑANA-YOGA

ONE OF THE FOUR PRINCIPLE YOGAS in Hinduism is 'jñana-yoga',
the way of the realization of God through the discriminating intellect. And the essence of jñana-yoga is expressed by the great Indian
sage Sri Ramana Maharshi of the holy mountain Arunachala by
means of a single question:

'*Who am I?*'

— to which the discriminating intellect answers:

I am not the body;
I am not the senses;
I am not the mind;
I am not even the sense of 'I'.

No. What I really am
Is the Eye observing all this,
Calmly observing even the sense of 'I'
As something other.

I am the *Atman*,
The Eye who sees all
But is never seen,
Because no eye can
See itself.

When one is enclosed within an artificial, humanly-created environment, it's hard to really see even that 'I am not the body.' After all, the chairs are dead, the appliances are dead, the cars are dead—my body, on the other hand, is alive; it's something more, something realer than the television, the elevator, the office equipment. I can feel it. It's really me. And when one is in an environment of other human beings who are all expressing their separate selfhoods—and that includes most business and social environments—the same rule applies: My body over here is me, as opposed to that body over there, which is her; the same goes for my thoughts and feelings. I may try to communicate my thoughts, express my feelings, or unite my body with the body over there, but that surely doesn't mean that this body, these feelings and these thoughts are no longer mine. Of course they are. Who else could they belong to? It's axiomatic.

But when one is in a natural, living environment, an environment that possesses life, like we do, but does not possess serious heavy ego, then one can begin to feel how one's body is a part of nature, part of the outer world, just one more living organism among the bugs and plants and birds...

And one's feelings too—they turn out to be continuous with the environment just like the body is, though with a subtler quality of that environment...

And one's thoughts—are they not simply the human way of what is, expressing itself to what is, just as the sound of the wind through a pine tree is particular to the pine tree and, though different from the sound of the wind through a eucalyptus tree, still, it's made by the same wind...?

And one's I-sense—is it any different, really, from the embryonic I-sense in the eye of that lizard, who knows who he is and knows that he doesn't want to die, just as that rock, lying where it lies, somehow intends to lie there, in perfect obedience, until moved by a superior force?

And isn't that I-sense simply the way that Infinite Life twists itself into multiple, particular existences, exactly the way that brilliant little green-gold bug has recently twisted it, for her own little purposes? It is not something that always happens, and always ends?

And isn't there Someone watching all this
at that rare moment when no-one is watching?

❧

XIX

NATURE AS SYMBOL

A SYMBOL IS SOMETHING that stands for something else—not one that is arbitrarily chosen, as if apples were randomly chosen to stand for oranges, but something that has an organic connection to the thing symbolized.

A symbol is a partial emanation of a higher, more integrated, more meaningful reality into a lower, less integrated, less meaningful world—which means that 'creation' and 'symbol-formation' are synonymous on all levels, from the writing of the text you are presently reading to the creation of universe. In the stunningly succinct words of René Guénon, 'the effect is a symbol of the cause', since, in the words of William Blake, 'every material effect has a spiritual cause.'

So all material forms symbolize higher spiritual realities—and this is especially true of virgin nature before humanity has interfered with it; of revealed sacred texts, such as the Vedas, the Torah and the Qur'an; of sacred art like the icon, the gothic cathedral, the traditional mosque or the Zen garden; and, to a lesser degree, of the traditional crafts. In all these forms, the symbol is a transparent manifestation of the reality symbolized.

As for other man-made objects and environments, they also symbolize spiritual realities, but realities of a lower order, often even a demonic one. Ugly office buildings, ostentatious advertisements, frivolous or degraded consumer products violate the human form as much as they pollute the natural environment. Symbols they are, because there is nothing that is not, but in a sense they are like 'symbols with an ego', designed to agitate or freeze the feelings, manipulate and darken the mind.

On the other hand, most undegenerate primitive or even pre-capitalist craft objects have an almost sacramental quality, because practical function has not yet been divorced from symbolic form. The products of such craftsmanship are at once useful tools, works of art, and ritual objects—this last because the symbolic meaning of the necessary actions of daily life has not yet been obscured. And the archetypes of most pre-industrial craft objects are to be found in Virgin Nature: plate, knife, net, basket, spear are leaf, sharp rock, bird's nest, spider's web, deer's antler. And all such natural forms and species are known, in spiritually-centered, traditional cultures, as 'words' of the Creator.

Traditional humanity reads nature like a book, knowing it as God's original scripture; and this is nowhere more explicit than in the early Fathers of the Church:

> As for those who are far from God. . . . God has made it possible to come near to the knowledge of him and his love for them through the medium of creatures. These he has produced, as the letters of the alphabet, so to speak, by his power and his wisdom. . . . (Evagrius of Pontus)

And:

> [The Logos], while hiding himself for our benefit in a mysterious way, in the *logoi*, shows himself to our minds to the extent of our ability to understand, through visible objects which act like letters of the alphabet, whole and complete both individually and when related together. (Maximos the Confessor).

This parallel between nature and scripture is so complete that Origen was able to say:

> we must necessarily believe that the person who is asking questions of Nature and the person who is asking questions of the Scriptures are bound to arrive at the same conclusions.

And if the natural world is a book, then a holy book may in some sense be the world of nature transposed to a different level. Thus a sacred text, rich with natural symbols, allows a kind of dialogue between nature and human language, in which the inspired consciousness of humanity is a world in the form of a book, and nature, as illumined by this inspiration, is a book in the form of a world.

From the Islamic perspective also, the natural world is a tapestry woven with the 'signs' of the Creator—the Arabic word for 'signs', *ayat*, being the same one used to denote the 'verses' of the Qur'an, thus making the correspondence between nature and scripture explicit. According to the Holy Qur'an:

> In your creation and in all the beasts scattered on the earth there are signs for people of true faith. In the alternation of night and day, and in the provision which Allah sendeth down from the heavens whereby he quickeneth the earth after its death, and in the distribution of the winds, are signs for people who are intelligent (Q. 45:4–6).

And:

> Truly the creation of the heavens and of the earth, and the succession of night and day, and in the ships which speed through the sea with what is useful to man, and in the waters which Allah sendeth down from the heavens...and in the order of the winds, and the clouds that run their appointed courses between heaven and earth, are signs indeed for people who are intelligent (Q. 2:164).

The Judeo-Christian expression of the same doctrine can be found in Origen:

> The apostle Paul teaches us that God's 'invisible nature' has been 'clearly perceived in the things that have been made' (Romans 1:20): what is not seen is perceived in what is seen. He shows us that this visible world contains teachings about the invisible world, and that this earth includes certain 'images of celestial realities'.... It could even be that God who made the human race 'in his own image and likeness' (Genesis 1:27) also gave to other creatures a likeness to certain celestial realities. Perhaps this resemblance is so detailed that even the grain of mustard seed, 'the smallest of seeds' (Matthew 13:31), has its counterpart

in the kingdom of heaven. If so, by that law of nature that makes it the smallest of seeds and yet capable of becoming larger than all the others and capable of sheltering in its branches the birds of the air, it would represent for us not a particular celestial reality but the kingdom of heaven as a whole. In this sense it is possible that other seeds of the earth likewise contain an analogy with celestial objects and are a sign of them. And if that is true for seeds it must be the same for plants. And if it is true for plants it cannot be otherwise for animals, birds, reptiles and four-footed beasts. . . . It may be granted that these creatures, seeds, plants, roots and animals, are undoubtedly at the service of humanity's physical needs. However, they include the shape and image of the invisible world, and they also have the task of elevating the soul and guiding it to the contemplation of celestial objects. Perhaps that is what the spokesman of Divine Wisdom means when he expresses himself in the words: 'It is he who gave me unerring knowledge of what exists, to know the structure of the world and the activity of the elements: the beginning and end and middle of times, the alternations of the solstices and the changes of the seasons, the cycles of the year and the constellations of the stars, the natures of animals and the tempers of wild beasts, the powers of spirits and the reasonings of men, the varieties of plants and virtues of roots; I learned both what is secret and what is manifest' (Wisdom 7:17–21). He shows thus, without any possible doubt, that everything that is seen is related to something hidden. That is to say that each visible reality is a symbol, and refers to an invisible reality to which it is related.

So every visible form is a sign of God; or, to say it another way, a name. My name is me, and not me; it is me in the vibrating air, in the ear and mind of another. When you've talked with a stranger long enough to ask him his name, and he has in fact pronounced that name, the sound 'John' reverberates from your inner hearing back to the face and form of the one who is now 'John'. It illuminates him; gives him depth; makes him realer to you—realer even to himself.

In just this way, the forms of Virgin Nature are words spoken by

God; and since That One is the only Reality, every word God speaks is necessarily one of That One's names—names like you, and me, and that shrieking bluejay, and the rising Sun. Since God is the only Reality, all things symbolize Him—and these symbols are unveiled at the moment when human consciousness encounters the world as it really is, with no intervening ego to separate them, since God is Lord equally over the object perceived and the subject perceiving it. As it says in the Qur'an:

> We shall show them our signs on the horizons and within them-selves until they are assured that this is the truth. Doth not thy Lord suffice thee, since he is over all things the Witness? (Q. 41:53)

To cultivate a vision of Virgin Nature as a tapestry of symbols, as the Creator's first book in which every form is a letter, every sentient being word, and every vital process a chapter, you must in a sense become profoundly naive; you must be guileless and simple-minded enough to know the kind of innocence which comes after experi-ence, not before it. The following question is an exercise in this extremely sophisticated form of simple-mindedness, the answer to which only seems complicated after it has been put into words:

Q: Why is the sky blue?

The scientific answer to this question involves the various wave-lengths of radiant energy within sunlight, the gaseous composition of the atmosphere, the laws of refraction, etc. But the metaphysical answer is: Because there is a higher world known as the angelic plane—in Sufi terms, the *malakut*—one of the earlier reflections of the Divine Nature, where all is high, cool, serene and clear, where perception is pristine and limpid, free from the agitation of discur-sive thought, where separate objects—like birds, the Sun, and the daytime Moon—are perfect expressions of their enveloping matrix, which is the essence of clear perception, the essence of Air. Given that this higher world exists in a relative eternity 'previous' to our experience of time, and that all realities radiate their particular qual-ities, reflection upon reflection, into ever more constricted and literal worlds, then that angelic world must, through a series of spiritual, psychic, subtle material, and finally physical processes, ultimately reveal itself to our physical eyes as the luminous, blue, overarching daytime sky.

The scientific answer to 'why is the sky blue?' is not 'wrong'; it's simply that it deals only with the solidified and already-decaying surface of things. And so to apply it to questions of ultimate origin is irrelevant, a fundamental mistake. As far as the ultimate nature of blue sky is concerned, the vision of a child who has been taught more in terms of religious than of scientific myth, who sees the beautiful, soft blue sky with its floating white clouds as 'heaven, the place where the angels live', is closer to the truth than the vision of the astronomer or meteorologist. (William Blake's *Songs of Innocence* were written to demonstrate precisely this.)

Using this same way of looking at things, we can answer other questions about the inner reality of nature:

Q: Why is the Sun hot, bright and radiant?

A: Because the Divine Intellect is a synthesis of Love (heat) and Knowledge (light), with absolutely no distinction between them, since to know God is to love Him, and to love Him is to know Him—and because God's Self-knowledge must necessarily radiate and illuminate all creatures, since those creatures are nothing in essence but particular forms of that Self-knowledge. As Dionysius the Areopagite says:

> What praise is not demanded by the blaze of the sun? For it is from the Good that its light comes, and it is itself an image of the Good. . . . I am certainly not asserting in the manner of the ancients that the sun actually governs the visible world as god and maker of the universe. But since the creation of the world, the invisible mysteries of God, thanks to his eternal power and godhead, are grasped by the intellect through the creatures.

Q: Why do trees branch?

A: Because, quickened by the sunlight of the Divine Intellect, the receptive Divine Substance that the Hindus call *Prakriti*—

hidden, like the roots of a tree, in the undifferentiation of its own nature—rises into manifestation, and ramifies, until it bursts into leaf and flower as the 'ten thousand things', all the forms of the visible universe.

Q: Why are eagles as they are?

A: Because the consciousness of creatures naturally aspires to rise and unite itself with the radiant Intellect that created it, like an eagle soaring into the Sun; and because inspiration from that Intellect descends upon creatures when and how it will, swiftly and unerringly, and assimilates them to its higher truth, as an eagle dives to devour its prey.

Q: Why are deer as they are?

A: Because the Divine Nature is filled with subtle, elusive energies by which God intuits the qualities of His own potential manifestation while they are still tender and embryonic, like deer sampling young weeds and spring grasses.

Q: Why does rain fall and water run down hill? And why does water evaporated by the Sun disappear?

A: Because the life-giving Truth of which the universe is made naturally descends the hierarchy of being from Source to manifestation in an increasingly visible manner, just as human creativity moves from cloudy intimation through clear conception to completed act, while the return of manifestation to its invisible Source, under the influence of Divine guidance, must, as it progresses, become less and less visible to outer eyes.

The word *maya* is related to the Sanskrit root meaning 'to measure'. The scientific worldview, based on measurement, ultimately draws us further into illusion, since it shows us the world not as it actually appears, but only an edited version, the world as we judge it to be based on thought and experiment. Symbolic consciousness, on the other hand, teaches us to see, again, the world as it actually appears to the full range of our perception. We are taught in school that ancient man believed, erroneously, that the Sun orbits the Earth, whereas the truth is that the Earth orbits the Sun. Ptolemy was wrong; Copernicus was right. But one of the implications of Einstein's theory of relativity is that the Earth's-eye view of the motion of the heavenly bodies and the Sun's-eye view are equally arbitrary, therefore equally valid. Why call one 'true' and the other 'false'? It all depends upon your point-of-view. However, in reality, neither view is arbitrary, since each has its own particular symbolic meaning. The heliocentric view demonstrates how the entire realm of manifestation—the 'Earth'—is contingent upon, a 'satellite' of, the eternally-creative Divine Intellect. But the geocentric view is, if anything, even more richly symbolic, precisely because the point-of-view is that of the human eye itself—a Sun's-eye view being possible only in the abstract because no human eye can exist on the Sun. Speaking in terms of effective symbolism, the way the cosmos actually manifests to our senses is the way it most truly is, since the more concrete our perception is, the richer it is in symbols addressed directly to us and 'speaking to our condition'. This is the famous correspondence between Macrocosm and Microcosm, of which Jung's theory of 'synchronicity' is a faint and flickering shadow.

Quantitative data bind our perception to their own level, which is the outer surface of things, at once material and abstract. Symbols, being both concrete and transparent, lead our perception beyond the symbolic level, all the way to the Thing Itself. As Clement of Alexandria (AD 140–c.220) says:

> By meditation....we are no longer considering the physical properties of an object, its dimensions, its thickness, length or breadth. What is left from now on is only a sign, a unity. . . .

Symbolic consciousness purifies the senses by breaking our attachment to sense-objects. It does so by transforming these objects from material facts into *truths*. It is possible to possess material objects or facts; it is impossible to possess truths, since they, of necessity, possess us. Blake: 'If the doors of perception were cleansed, everything would appear to man as it is, infinite.' Contemplating the forms of the world as symbols, we contemplate God—but we need always to remember that God is more than an Object to be looked at: He is looking too—and since, in simple truth, That One is the Only Being, our contemplation of Him is actually His contemplation of Himself. As Maximos the Confessor says:

> If invisible things are seen by means of the visible, the visible things are perceived in far greater measure through the invisible by those who devote themselves to contemplation. For the symbolic contemplation of spiritual things by means of the visible is nothing other than the understanding in the Spirit of visible things by means of the invisible.

Or, in the words of Meister Eckhart:

> The eye through which I see God, and the eye through which God sees me, are the same eye.

⊚

XX

REALITY AND NAME

ANCIENT WISDOM-TEACHINGS from cultures and religions the world over say something like this: 'Reality knows itself, and the form of this knowledge is Humanity.' Many myths tell the story of the Universal Man or Primordial Adam, an ancient King or original Giant, the first created being, who later died, or was sacrificed, or fell asleep, after which he was unfolded, or broken up, into all the forms of the Universe. In the words of William Blake: 'Albion anciently contained in his mighty limbs all things in heaven and earth—but now the starry heavens are fled from the mighty limbs of Albion.'

What does this mean? It means that all the forms of nature are the scattered limbs and organs of a vast Human Form. But what does that mean?

To answer, we need to start from the two ends of the question: God, and man; Absolute Reality and human experience. From God's side, we can say that Absolute Reality must know itself, otherwise it wouldn't really be Absolute, there would be a hole in it somewhere, a hole made of darkness, of ignorance. And yet, from our own experience, we know, 1) that we can't know ourselves without having a view of ourselves, and 2) that every view of ourselves is always partly right and partly wrong; it emphasizes certain aspects of who we are, and leaves others out. So the only way to know ourselves perfectly would be to stop generating partial views of ourselves and simply be ourselves, exactly as we are. But that perfect self-being would also have to be a perfect self-knowing, otherwise it would be unconscious of itself, and thus not really perfect. Such perfect self-knowledge would necessarily be a simultaneous vision of all the possible versions of who we are in a perfect synthesis—a unity made up of an infinite number of views of who we are, perfectly united with the *reality* of who we are.

If we were self-subsistent entities, determined by nothing beyond ourselves, this would be our natural way of being. But that's not the way we are; we only come into being at the command of the Creator. He alone is self-subsistent; He alone possesses absolute Being perfectly united with absolute Knowledge. In Sufi terms, His Absolute Unity (*ahadiyya*) is perfectly at one with his Synthetic Unity (*wahadiyya*)—the unity of His Names, which is the infinity of all possible views or versions of Who He Is. His Synthetic Unity is *ALLAH* ('the Deity',); His Absolute Unity is *HU* ('He').

Only God can perfectly know and perfectly be Himself, since only in Him are Knowledge and Being perfectly united. Creatures like you and me can only partially know ourselves and so only imperfectly be ourselves; or it can be said with equal accuracy that we can only imperfectly be ourselves and therefore only partially know ourselves. The incomplete self-knowings of creatures are necessarily partial because they are all *parts* of God's complete self-knowing, while our self-being is imperfect because Being is truly possessed by God alone; all we possess, as creatures, is nothingness. God absolutely possesses all the views, of whatever object, which are relatively possessed by creatures, from the tiniest elemental being, through the plant, animal, human, animic and imaginal kingdoms, and the angelic choirs, to the highest Seraph. And He knows them all as views of Himself, since He is the only Being. The creatures below Humanity (and below the Jinn, denizens of the animic plane, who, like us, also possess free will) are fixed in their views, which veil God from them (though not them from God); the creatures above Humanity, the Angels—who made their free choice to accept God in eternity—are also fixed in their views, which allow each angel to contemplate God by means of the single Idea he knows and represents (though not as He contemplates Himself). But Humanity alone—here defined as any race of creatures occupying space and time but cognizant of eternity, with a will that is free to choose or reject God—possesses the power to choose God by sacrificing their views: views of other creatures, views of themselves, and ultimately views of the Deity.

Here begins man's side of the question, the empirical side, which starts with the understanding that we as human beings necessarily generate alternative views of ourselves and the world around us, while animals do not seem to, at least not to the same degree. The alternative views seem to be based almost entirely on language, with its ability—for good or ill—to *fix* and *train* perception. (Human language operates through symbols, which transcend time; the rudimentary languages of animals have nothing in them beyond a response, though often a very sophisticated one, to the situation at hand.) But if we know what the spiritual Path really is, we will understand that human beings also have the power to dissolve their views of themselves and the world, and let both self and world be only as God knows them to be—in other words, as they really are.

To sacrifice our views of ourselves and the world is to return to the Absolute Reality that exists before such views are born. But once we have returned to that Reality for good, then we can generate views without getting trapped in them, without thinking that a particular view of the world *is* the world, or that a particular self-image is our real self. Furthermore, since these views are no longer confused with the Thing Itself, they appear as direct and spontaneous expressions of the Thing Itself: 'Sangsara is Nirvana'.

To know that our views of Reality are not Reality Itself is to realize the transcendence of God. To know that our views of Reality are symbolic renditions of Reality Itself is to realize the immanence of God. To know that the views spontaneously generated by Reality are ultimately identical with Reality is to realize the 'tantric' synthesis of transcendence and immanence, where the polarity between Source and Manifestation is dissolved in Union. Each of these three 'stations of wisdom' is greater than the last; nonetheless the greater view does not annihilate the lesser, but contains it: without transcendence, no immanence; without immanence, no Union.

So we, as human beings, are capable of both generating views and going beyond them, which is another way of saying that the universe is created, and destroyed, through humanity—not humanity in our time-bound material identity, but the eternal human function within the Divine Nature, which both creates us and operates through us. As we in our little selves create and transcend our views of things, we participate in the creation and reintegration of the universe via our eternal archetype, the Divine Humanity, who is the totality of God's knowledge of God.

The universe is eternally created because Being is always emanating a view or symbol of itself by which it can be known. It is eternally destroyed because the symbol, no matter how accurate it is, is not the Thing Itself. It is eternally beyond creation and destruction in the depths of the Divine Nature—a truth that Frithjof Schuon calls '*maya-in-divinis*'—because the symbol is of one essence with the Thing Symbolized: If God is the only Being, then appearances are none other than the Real. Therefore, since God eternally creates and destroys the universe through us (in a sense), the Nature we see around us is really our 'emanation', our *shakti*, which is the inner meaning of the myth that Eve was created from Adam's rib while he slept—in other words, at the exact point where (undoubtedly via human language) he began to mistake his view of Reality for Reality Itself. The myth that Eve was created from Adam does not mean, however, that Adam, as a man, was somehow realer than Eve, a woman, because the truth is that there was no Adam until Eve was separated; the 'Adam' created by God and placed in the garden was not of the male sex, but an eternal Androgyne, the Primordial Humanity, in which experience and experiencer, self and world were not yet divided (cf. Plato's Symposium). This is indicated in Genesis as well, where it is said that 'God created man in his own image....male and female created he them.' (Genesis 1:27)

Subject and object, in other words, are born together: until we have a view of the world, we can have no view of ourselves; until we have a view of ourselves, we can have no view of the world. So when we look at Nature (whether 'natural' or man-made) we are seeing the other half of ourselves, the part that was exiled from us when we 'ate of the Tree of the Knowledge of Good and Evil' and fell into the subject/object duality, into matter, energy, space and time. But it is easier to really see this in Virgin Nature, since when we are in the presence of man-made 'nature', the lesser truth that what we see is based on our secondary ego-intent obscures the greater truth that even cars and buildings ultimately 'arise of themselves', whereas Virgin Nature is clearly other than us, obviously not a creation and slave of our ego-intent, since, in terms of time, it came before us. Only what is undeniably other than the ego-bound self can be recognized as the lost half of our human soul, of that integral, androgynous being we were (and are) before time began.

The Old Testament says that *Adam named the animals*. The Qur'an maintains that *before he was materially created on earth, Adam, by God's command, told the angels their names*. The animals, in other words, are earthly manifestations of celestial archetypes, whereas the angels symbolize the archetypal Names of God, which appear as animals to our earthly eyes. And so the angels are both the names of the animals and the Names of God; they are the invisible *logoi* (words) through which God creates all the visible forms of the universe.

On the other hand, the Buddhists say: 'To name something is to kill it.' This certainly sounds like a contradiction, or at least a difference of opinion—but is it?

Adam did not name the animals on an arbitrary basis. He could name them because he *already knew their names*. He knew them because they were the innumerable versions of Reality or Names of God of which he, being God's central act of self-understanding, was composed. They were his limbs and organs; as he named them, they appeared in the outer world. Through the medium of the Primordial Adam, God's Word unveiled the Divine Names, and cosmos— harmonious order—was born.

To name something is to create it by separating it from all that is other than what is contained in that name, as in our phrase 'to make something out'. It is to distinguish it from its background, to see it as a separate entity. Naming is like hunting: We delineate something as a form, transfix it with a name, and eat it for knowledge. Our paleolithic ancestors apparently did the same thing; the walls of their caves suggest that they first drew a picture of the quarry, then hurled their weapons at this *imagined* beast. And since written words and letters began as pictures, it would be good poetic logic to imagine them shouting the beast's *name* as their weapons struck stone—a name which, for the purposes of the 'real' hunt, would then become a silent invocation. All of us, as children, created our world of separate yet related objects in just this way: by uniting imagination and sense perception through the medium of language.

Our sense-experience of the world around us is conditioned by two things: the structure of our nervous system, and our learned patterns of perception: nature and nurture. Dogs can hear sounds that are too high for us to 'make out'; insects can see ultraviolet light that is invisible to us. And as many studies in the psychology of perception have proved, the way we see the world is largely determined by how our perception is trained, by both our culture and our individual experience. And this patterning of perception happens to a great extent through language.

But only human beings can know that the world-as-perceived is not the totality of the world-as-it-is, because only we can deliberately alter our patterns of perception. An animal is imbedded in its own genetically-determined perceptual pattern. And even if this pattern is altered by training, the animal cannot deliberately choose to alter it, or even know it has been altered. Only the human being, then, is an open door leading beyond the universe-as-perceived to Reality as it is—which is why animals sometimes look at us with a mixture of hope and shame; they feel that we represent a greater good beyond their understanding. (Poet Gary Snyder says somewhere that animals are often fascinated by humans; they see us as very beautiful beings.) Nonetheless, how many of us really understand:

1) That the way we see the Universe is based on a pattern of perception, that it is not an absolute but a relative view of an absolute, objectively true but inevitably limited: a Name;

2) That our view of the Universe always changes, but unless we learn how to change it deliberately, those changes remain unconscious;

3) How to actually change that view, so as to experimentally demonstrate that all views of the Universe are relative;

4) That there is an Absolute Reality, manifest in all views but beyond all views;

5) That there is a Path through which this Absolute Reality can be realized.

In other words, how many of us are actual, not merely virtual, human beings?

The Qur'an mentions something called 'the Trust' which God 'offered . . . unto the heavens and the earth and the hills, but they shrank from bearing it and were afraid of it. And man assumed it' (Q.23:72). (We can imagine the Big Bang and the expanding universe as the heavens and the earth and the hills, the elemental energies of the cosmos, fleeing the Trust.) This Trust is nothing but our ability to know and do the above five things, and so function as a way Out, a Door leading from things-as-they-seem to things-as-they-are (after which we can say, without entanglement in the material surface of things, that 'things are exactly what they seem'). If we do not fulfill this Trust, then we are not yet, or no longer, human; we are nothing but animals without the protection of a single unchanging pattern of instinct and perception, chaotic animals who are always shifting from pattern to pattern without being able to control ourselves, who are addicted to 'trying things out' and so end up destroying everything they touch.

The way out of the environmental crisis, then, is not for us to become more 'natural', more like fish and birds and flies, but for us to become more human. Our duty is not to manipulate nature, beyond the minimum necessary for our survival, but to *contemplate* it as a symbol of our Creator, thus allowing it to transcend itself— after which we will see just how much and what kind of intervention will enhance the natural world instead of destroying it. We are given 'dominion over the fish of the sea and the fowls of the air and over every living thing that moveth on the earth' only in this sense, the sense conveyed by William Blake when he said: 'Where man is not, nature is barren.'

۞

Above, I quoted the Buddhist proverb, 'To name something is to kill it.' This simply means that to separate something from its background by means of language is to fracture the original Unity. But if we can learn to witness this process in complete detachment, then we are not fracturing the original Unity, but unfolding it into manifestation. And it is not we who are really doing it, but God's Word speaking through us.

When you are walking in the woods, to know the names of the plants and animals and minerals makes them realer to you; they become friends of yours; you know them by name. Of the other hand, if you walk along surrounded by a cloud of your verbal definitions of things, you are seeing the world through a haze. (Have you ever noticed how some people have to keep talking when they are in a natural environment, to distance themselves from that hidden Other Side of their being?) So one of the arts of appreciating Nature is to know the names of things without thereby starting to act like a professional biologist to whom Nature is nothing but shop-talk—AND to be able to let go of verbal definitions instantly, so that everything merges back into the original Unity. (When WE are naming things, we rely on our eyes, since we are in the process 'making things out'; to return to the original Unity, we need to shift some of the burden of perception to our ears, and listen to God naming things. Listening softens the gaze, till it can just make out the Nature Spirits behind the rustling boughs. And if we listen perfectly—that is, if we stop talking to ourselves completely—then the Eye behind the eye may open by itself, and let us see into the Heart of things.)

۞

XXI

LISTENING
TO THE EARTH

THE WORLD IS GOD'S BOOK, in which every form is a letter or sentence. But the world is also God's spoken Word, in which every sound is a reverberation of the Logos, by which, according to the Qur'an, God has only to say to each possible thing 'Be!', and it is. In the words of the Gospel of John, 'In the beginning was the Word , , , and without him was not anything made that was made.'

The sense of sight is symbolically related to the faculty of discrimination, and also to explicit knowledge. 'Seeing is believing'; the forms of stationary objects appear established and unchanging; they have a factual quality. And there is a way in which what we see is cut to the size of the seer; it is a pattern which, unconsciously though somehow intentionally, we put together or 'make out'. This does not mean that the material world is only a projection of our subjectivity, since the process by which we pattern that world is as truly and objectively there as any granite boulder. But still, the world we see is in some sense our 'twin;' the one looking and the object of his or her sight somehow 'see eye to eye'.

The sense of hearing is different; it is related, symbolically, more to the continuous act of creation than to the catalogue of what has already been created. (From an opposite though complementary standpoint, sounds can also represent the realm of contingency, since they are immersed in time, whereas visible objects can symbolize the permanent archetypes or Platonic ideas; but that's not the view, or word, we are considering here.) Sounds are like objects that haven't solidified yet, or like the eternal prototypes of which material objects are the outward expression. Sound is bigger than us; it surrounds us and washes over us. We can intentionally look in a particular direction, but we can't listen in a particular direction. Sounds simply come to us, unpredictably, inexorably, from beyond what we know. This is why the act of hearing is organically related to the act of obedience; instead of judging and discriminating, we simply 'hear and obey'. With our eyes we investigate, we spy things out; but the knowledge that flows into our ears is a gift, not an acquisition.

The flow of the Tao—in western terms, the will of God—comes into our experience through the dimension of time, which is why the ancient Chinese book on the wisdom of the Tao is called 'The Book of Changes' or *I Ching*. We become sensitive to the will of God by becoming intimate with time, and one of the best ways to do this is by listening. If we listen deeply enough we can hear not only the changes of time moving through conditions, but the resonance of their eternal Source, the Logos sounding behind them.

So listening can be a form of contemplation. There are, say, four levels to this contemplation, depending upon the depth to which, and from which, we listen. These are: listening to the sounds of the world; listening to the inner sound in the brain center; listening with the heart center; and listening with one's whole being. In terms of the spiritual practice of reciting a Name of God, which the Hindus call *japam*, the Sufis *dhikr*, and the Eastern Orthodox Christians the 'Jesus prayer', listening to the sounds of the world corresponds to vocalizing the Name, or (in the Hindu practice) 'hearing all sounds as *mantra*'; listening to the inner sound corresponds to the silent recitation of the Name; listening with the heart center, to the experience of God speaking His own Name within us; and listening with one's whole being, to witnessing all events, in the world or in ourselves, as acts of God.

In listening to the sounds of the world, you simply sit and attend to all the sounds within your range—birds, wind in the trees, flowing water, traffic sounds, human voices—hearing them as the voice of the Deity, the vibration of the primal creative Source of the

Universe, finally reaching your ears. As you listen to the sounds of the world, you will realize that the lapses in your attention correspond to obsessive sub-vocal speech—and the way to quiet this mechanical chatter is to listen to the inner sound in the brain center, that hum or hiss or high-pitched whine we usually hear when no sounds are coming in from the outside, the 'sound of silence'. If the sounds of the world are like the leaves and branches of a tree, the inner sound is like the trunk.

In listening with the heart center, you place your attention within your chest, and attend to the soundless seed of your own being, the point where your body and mind are continually emanating from Source, where God is eternally creating and re-creating you, and the world around you, instant-by-instant. If the sounds of the world are like the leaves of a tree, and the inner sound in the brain center like the trunk, then the soundless resonance in the heart center is like the root. At the heart center you are listening not to inner or outer sounds, but to subtle feeling-tones that are also direct, immediate 'knowings'.

In listening with one's whole being, one totally surrenders to God, which entails the total annihilation of the listener. If listening to the sounds of the world corresponds to the leaves of the tree, listening to the inner sound to the trunk, and listening with the heart center to the root, then listening with one's whole being is like the Ground.

When you listen to the sounds of the world, you are part of the world around you, a universe created by God before you were born, bigger than you in space and older than you in time: the object experienced appears to be the source of the experiencer.

When you listen to the inner sound in the brain center, all the sounds and forms of the world are experienced as the 'echoes' of that sound, which is the sound of God creating, *through you*, the universe you perceive: the experiencer appears to be the channel or immediate source of the object experienced.

When you listen with the heart center, you understand how you and the world are created by God together, simultaneously, in eternity, from the same Heart: experiencer and object experienced are united in the state of the original Androgyne, the Primordial Humanity.

When you listen with all you are, then there is no experiencer, no object of experience, no self, no world, but God alone. (NOTE: These practices, since they are not part of an integral spiritual Path, but only words in a book you are reading, may give intimations of these levels of being, but not stable realizations of them—unless God wills otherwise.)

To really listen to the sounds of the world, your attention needs to be detached from the world and operating on a deeper level. The eye entangled with sight is blinded; the ear entangled with sound is made deaf. The same is true of the inner sound and the soundless resonance of the heart: in order to hear them clearly, you need to be 'behind' them. The sounds of the world are 'listened to' by the inner sound; the inner sound and the sounds of the world are listened to by the Heart; the Heart, the inner sound and the sounds of the world are listened to by the Ground of Being. And when the Ground of Being speaks for itself, and as Itself, then there may be sound, but there is no listening: the act of attention is dissolved.

The four levels of contemplative listening can be practiced separately, but the best way is to synthesize them, to simultaneously listen to the sounds of the world, the inner sound and the resonance of the heart from the standpoint of the Ground of Being, thus demonstrating the Mahayana Buddhist doctrine that 'sangsara is Nirvana', or the Sufi teaching that 'the relative is the bridge to the Real'. And the best place to practice this absolute listening is in a wooded area, during a gentle wind. (From the standpoint of the outer ear or the inner sound or the Heart, it seems that 'you' are listening. But from the standpoint of the Ground of Being, only God is listening, to the speech of no-one but HE: in Him, the act of speech and the act of listening are One.)

The use of the image of the tree to illustrate the four levels of contemplative listening is not arbitrary (though the sound of a running stream or the pounding of ocean waves might just as well have been chosen), because true symbols always have an organic relationship to the thing symbolized. The symbol is what takes us beyond the literal level of things—or, to say it another way, once the literal level is transcended, the symbol appears: 'The letter killeth, but the spirit giveth life.' If we think that the Sun is God, or the wind is the Holy Spirit, then we are literalists, idolators. But as soon as we realize that the Sun is not literally God, only a ball of incandescent gas, that the wind is not literally the Holy Spirit, only a mass of air stirred by the heat of the Sun, then we have a chance to see the Sun as a symbol of the Divine Intellect, and the wind as a symbol of the Divine Power—and, furthermore, that Sun and wind, as symbols, are also *examples*, actual instances, of the operation of Divine Intellect and Divine Power in this world. As it says in the Qur'an, *God is the light of the heavens and the earth*. So when we sit under a tree, listening in contemplation to the sound of the wind, we can know that, although this tree on the literal level is nothing but a vegetative life-form, sprouted from seed and destined to rot back into the soil, on the symbolic level, and thus in its essential reality, it is that Tree whose leaves and branches are the Universe, and whose roots are buried in the invisible Source of all—the Tree of Life itself.

XXII

NOSTALGIA
AND RETURN

POET ROBERT BLY SAYS SOMEWHERE that the central feeling-tone
we experience when we relate to the natural world is *melancholy*.
And (central or not) I know that melancholy well; I'm sure most of
us have felt it. It's what the Germans call *weltschmerz*, 'world-
sadness'—a deep, sometimes heart-rending nostalgia, best captured
I think by the Native American music of the Andes. Part of it may
have to do with the great suffering the Earth is experiencing right
now; part of it may relate to the alienation from nature we feel, due
to the fact that most of us live in cities—but a great deal of it simply
has to do with what it's like to live in a material world. Listening to
the spine-tingling music of the Quechua Indians, descendants of
the Inca, or sitting in a mountainous region, where ridge after ridge
of snowy peaks fade back into the purple haze, you may suddenly be
struck by the sensation that the Earth is immensely old, and that the
burden of her age... is sorrow. How long she has been here! And
how long we have been here with her—much longer, perhaps, than
science can presently demonstrate. Generation after generation,
millennium after millennium of the human dead lie hidden in her
hills. How much love, and heroism, and labor, and sorrow have
been raised, and slain, and forgotten in her. Sitting among such
mountains it may slowly dawn upon you that looking into great
distances of space is the same as gazing deep into time. Those peaks
you see, rising, now, before your eyes could just as well be the peaks
of Atlantis, 100,000 human years in the distance.

But as we all know, or should know, nostalgia can never be requited, at least not in its own terms. Space is always expanding, time always moving away from the radiant Center of Being… and all those who follow them only become lost in the outer darkness. The nostalgic melancholy we experience when sitting alone on the ancient, turning Earth is not really the desire to be reunited with the ancestral dead; that's only the shadow of it. The feeling of 'ancientness' is not a sense of the past, though it may seem to be, but rather an intuition of the Eternal, as when God, in the Old Testament, is called 'The Ancient of Days'. What we actually long for in moments of nostalgia is Eternity, the deep well of the present moment, where all past generations, the loved ones we have lost, 'the years the locust hath eaten', are alive here and now as the permanent archetypes of all experience, the Names of God, forever bursting into dimensional existence from the Night of the Unseen.

The Qur'an says, 'All is perishing except His Face.' The implication is that in order to witness that Face we must let go of all that is perishing, simply let it die. And this is certainly true. But if we can really make this sacrifice, if we succeed in releasing the moments of our life into the flow of passing time, and thus in contemplating the Face of Him who does not perish, then, in a sense, the flow of time reverses. The dead are resurrected. The ancestors return. Time no longer *passes*—it *arrives*. And when this reversal is firmly established, then nostalgia is no longer a tragic and unrequited longing for the past, decaying moment by moment into ever-increasing stagnation and loss, the state Blake was talking about when he said 'Memory is Eternal Death', but a continually freshening and always deepening intuition of Eternal Life, where time is not forever slipping away into memory, but forever returning to God, who is alive in the Eternal Now. In Blake's words:

He who binds to himself a joy
Does the winged life destroy
But he who kisses the joy as it flies
Lives in eternity's sun rise

To practice the contemplation of nature as symbol is to reverse the flow of time. The symbolic vision of the natural world dawns at the exact point where nostalgia for the past turns, and enters the gravitational field of God in Eternity—that being the point of spiritual contemplation. In the words of the Qur'an: 'To Him does the whole matter revert.'

XXIII

'WE MUST SAVE THE EARTH'

YES, WE MUST; whether we will or not is another matter. Yes, massive efforts will be required, a veritable war for the life of the planet, and the human race. Nothing less will suffice.

And yet there is something in the quality of 'war', of 'massive efforts', where time takes over everything because time is always short, that is not entirely in line with the energies of fertility, of life. Pregnancy cannot be hastened; the seasons cannot be rushed. Haste is definitely required—and yet 'all haste is of the devil'. There is much to do, and quickly—and yet 'doing ends in death'.

How do we solve this apparent paradox? The *Tao Te Ching* says:

> *I do nothing and the people are redeemed.*
> *I remain at peace and the people forget their cunning.*
> *I avoid action and the people accumulate wealth.*
> *I want nothing and the people return to the simplicity*
>   *of their original nature.*

The speaker here is the 'sage', who is both the realized saint and the sage part of ourselves, the one who 'does nothing' because he realizes that only God is the Doer. Only the sort of action that springs out of this profound and receptive non-action has the power to restore balance rather than further upsetting it.

Until fairly recently, Western culture was naively proud of its ability to 'conquer' nature; among fools, such rhetoric is still sometimes heard even today. But if we are, or were, so confident of our ability to conquer, then why have we despaired of our ability to manage? After an empire is conquered, it needs to be administered—unless, that is, the empire was a civilized cosmos and its conquerors mere barbarians, who are at home in a state of relative chaos. If nature is inherently a less integral reality than commercial/technological man, and so destined to be ruled by him, then why have we given up trying to rule it? Why, for example, have we all but resigned ourselves to the eventual destruction of the Earth's ability to produce breathable air? If the managers of an automobile assembly plant were to decide to melt down the plant's machinery to make parts, or sell it off to cover wages and stock dividends, this could in no way be called 'efficiency'. Capitalism ought at least to be able to see the ecosystem as capital, not profit. But once we lose the vision of nature as a manifestation of the Deity, there is no end, apparently, to the circles of intellectual darkness through which we fall.

The triumphal atheistic materialism of modern societies is often called 'Promethean', after the titan who stole fire from Zeus, and defied him. But our fate appears, in some ways, to be more like that of another titan—Atlas. If we insist on conquering the Earth, what more fitting punishment could be imagined than for us to be condemned to carry the Earth until she crushes us—or until we remember that, in reality, none but the hand of the Deity has ever held her?

☺

Heraclitus said: 'The way up and the way down are one and the same.' What does this mean? That the spiritual Path, the way of return to God, is an exact reversal, or mirror-image, of the Way God created (or, to be strictly accurate, 'creates') the Universe. As manifestation unfolds, the Source of that manifestation is veiled. When manifestation folds up again, then that veil is rent (as at Christ's crucifixion) and its hidden Source revealed. This revealing-of-Source is also its re-veiling, since manifestation obscures what is revealed, while this disappearance is also a revelation, since it demonstrates that Reality in Itself is greater than any version of it. In hiding Himself, God reveals Himself through the Universe, which both hides and reveals His Essence. In the words of Maximos the Confessor:

the forms of visible things are like the clothing, and the ideas according to which they were created are like the flesh. The former conceal, the latter reveal. For the universal creator and lawmaker, the Word, both hides himself in his self-revelation and reveals himself in his hiding of himself.

According to the Hebrew Kabbalah, God creates the universe by a process known as *Tsim-Tsum*, the withdrawal of the Creator into His own hidden nature so as to compassionately 'make room' for the manifestation of those latent possibilities within His nature that are longing to be born. The image is that of creation by means of a self-humbling of the Divine Nature, a sort of courteous discretion, like that of a parent who unassumingly fosters those potentials within his children that lead them to become who they most truly are, who sets clear external limits, while guiding in secret. Through *Tsim-Tsum*, God humbles Himself to create the Universe, just as Christ humbled Himself to redeem it (cf. Philippians 2:5–9).

So creation is the manifestation of the hidden Creator; and the synthesis and focal center of this creation is the Human Form. This is why the 'way up', the reversal of God's creative act, the folding up of the cosmos happens through integral human consciousness alone, by means of self-purification, self-transcendence, and self-annihilation. When cosmic manifestation arrives at Humanity—a state fully realized only in the saints—the process of creation is complete. As soon as it is complete, it begins to transcend itself, to flow back into the unseen worlds, until it returns to the invisible Source of all.

From a deeper perspective, however, creation and return are simultaneous. God is always manifesting the universe, the universe always returning to God, because That One, as it says in the Qur'an, is both the Inner and the Outer, both the First and the Last; the Divine Nature comprehends, and also transcends, both the manifestation of the universe and its annihilation.

And so, given that this is the way things are, what are the implications of this metaphysical principle for the work of saving the Earth? To answer this question, we first need to take a look at the present state of 'the World', the collective mindset.

Contemporary society is disjointed, to say the least. People with a neolithic or even paleolithic cultural background—like, say, the Balinese and the Australian Aborigines—are thrown into the same global whirlpool with those whose ancestors have been civilized for thousands of years. And the global whirlpool itself, the maelstrom of industrialism and the post-modern 'information culture' is, as it were, actually composed of such disjunctures; they are its essential texture. Postmodernism as a cultural program is little more than an attempt to enshrine this fragmentation, to place the center of things on the jagged edge between one set of assumptions, one ethos, one belief-system, one aesthetic and another one with which it has no organic relation, while at the same time denying that such a thing as a 'center' can exist. The global economy is continuing to destroy the craft traditions and the viability of unskilled labor and labor-intensive farming, effectively outlawing subsistence lifestyles almost everywhere on the planet, while at the same time producing more and more information and computer-related jobs, the result being that those who are not temperamentally fitted to be manipulators of information are increasingly marginalized.

In terms of the quality of human consciousness, the most perfect expressions of this fragmented global culture are the electronic media, including computers and the 'cyberspace' they create, which might almost be defined as 'tools for mass training in the fragmentation of human consciousness and the numbing of the emotions, leading to an addiction to chaos.' We get to where we need fragmentation and disjunction in our view of ourselves, the world and other people, to such a degree that if we are suddenly deprived of the media-irritants we are addicted to, or confronted with an expression of organic wholeness, either in an object of art, in virgin nature, or through meeting with an integrated human personality, we experience the actual pain and fear of withdrawal—which, following the precisely applicable metaphor of drug addiction, is a confrontation with the damage the addictive substance has been doing all along, under cover of the glamorous intoxication we seek.

One consequence of this development is that many of us, especially those of us working in the 'white collar' world, feel too 'ethereal'. We don't have our feet on the ground; we don't experience our natural, physical human nature. After eight hours at the computer screen we're hardly aware of our own bodies, much less the world around us. We're too cerebral. We don't feel real.

One way we react to this imbalance is by trying to live our lives entirely on the mental plane, through either obsessive thought or obsessive fantasy, or both. And a further consequence of this imbalanced lifestyle is that we *idealize* the world of the senses that our minds are too frozen and agitated to fully experience.

The effect of this state of affairs on the possibility of a solid understanding of metaphysical principles is disastrous. The disaster has a specific form and progression, as follows: Our Promethean-materialist mindset produces a form of technological society that destroys both the natural world and anything deserving the name of human consciousness; but while our basic assumptions remain consciously or unconsciously materialistic, the form of awareness that our technological, information-based society enforces is ethereal, ungrounded and abstract. Its very 'subtlety' is a product of its essential grossness. It seems perceptive, but is actually sterile and empty. We are led to identify 'spirituality' with this kind of gross subtlety, this effete, hyper-cerebral consciousness, producing among other consequences the religious movement known as the 'New Age'.

Then we look around us and see that the natural environment is profoundly threatened. Forests are disappearing, species becoming extinct. We identify this devastation, understandably enough, with the imbalances we feel in our own souls; and such identification is, in fact, justified. But the *philosophical interpretation* we put on the relationship between human psychic imbalances and the destruction of the environment—*conditioned as it is by those very imbalances*—is so profoundly wrong that one wonders if the Devil himself could have devised such a finely-crafted error; or whether, in fact, he actually did.

The essence of this error is to believe that 'otherworldliness' is the cause of the environmental crisis, that nature is being destroyed because humanity *isn't material enough*. Numbed by our electronic technology, and driven nine-tenths out of our bodies by the stresses of modern life, we inhabit a cold, shrilling hell of phantom subtleties, where our environment is increasingly composed not of the natural world, or of the society of other people, or even of machines or buildings, but rather of our own nervous system, plus its global electronic extensions. (The socially 'well-adjusted' individual of the near future may well be the isolated, unmarried 'telecommuter' working out of his or her house or apartment, or trailer, compliments of the personal computer.) Consequently, we dream of grossness as if it were a lost paradise. We long to roll in the mud like pigs, if only it will return us to the bodies we are alienated from and the natural world we have lost. We dream of falling back, for rest and security, on the wide maternal breast of that world, only to read in the daily news how that breast is now riddled with cancer.

But no matter how insecure the material world becomes, our attachment to matter remains strong, and even increases, largely because contemporary materialist beliefs portray it as the only reality, and war against any philosophy that might remind us that matter is the creation not the Creator, that creation is ephemeral by nature since it is ruled by entropy, and that safety and stability are with the permanent archetypes alone, and their indestructible Divine Source. Profoundly threatened by social and environmental breakdown, we long for any basis of security—and then run to look for it in philosophies that make change paramount, and call eternity an illusion. We jump into the flood-swollen river of chaotic change, looking for a firm foothold. In our despair, we are fascinated by, and driven to worship, the very things that are destroying us. In terms of the environmental crisis, we deny the pitifully obvious—that the destruction of the natural world is the result of materialism—and try to make ourselves as gross and material as possible, to crush out all sense of the Hierarchy of Being, of the reality of the Transcendent, of the Divine Creator of all life. We think that if we squash ourselves down into matter, if we dissipate and materialize ourselves enough, we will somehow restore the natural world.

But this is not true. The more bogged in materialism we get, the faster we will destroy the Earth, a truth made daily clearer to everyone—except those who will not see. Materialism as a psychic fixation also distorts the physical experience of our humanity, since it hides from us the truth that the living human person is in reality a polar resonance between matter and Spirit—a truth that can be known only in contemplative Silence: We return to our bodies and the natural world not by idealizing and grasping after sense experience, but by quieting the mind.

The principle, again, is the Kabbalistic *Tsim-Tsum*: that God manifests the universe not by hurling Himself out of Himself—an act that is impossible to Him, since there is no place that is not He—but by withdrawing into His hidden Essence to 'make room' for it, since it is only by veiling His glory that He can avoid annihilating all manifestation in His greater Reality. The contemplative does the same. By withdrawing from manifestation on the 'way up', he leaves room for the Divine Nature to manifest through him on the 'way down', since our annihilation is God's revelation.

A story is told of the classical Sufi master Dhu'l Nun, an Egyptian Black who is believed to have introduced Egyptian/Hermetic lore into Sufism: There was a serious drought in Egypt; the Nile was not rising; famine threatened. The people ran to Dhu'l Nun, their greatest saint, to implore him to intercede on their behalf, and pray for rain. So Dhu'l Nun made an investigation into the Unseen to determine the cause of the drought, and saw that the only barrier to the falling of the rain—was he himself. So he left the land of Egypt, and the rain came.

There is really no other way. Out of the self-sacrifice of Jesus on the cross came an eternal Way of Salvation and a spiritual culture that has lasted two thousand years, and taught us to value works of mercy. Out of the Buddha's insight into the ultimate non-reality of self came an eternal Way of Enlightenment, and a spiritual culture lasting twenty-five hundred years, teaching the values of compassion and non-killing. Out of the *nirvikalpa-samadhi* of Ramakrishna came, under his successor Vivekananda, the hospitals, schools and wide-spread social service of the Ramakrishna Order. Out of the martyrdom of al-Hallaj came the power of Sufism to coexist with, and inwardly renew, exoteric Islam for over a thousand years. Humanity does not renew the world of manifestation by drowning in it, but by returning to the Source which created it, thus leaving an open doorway, empty of ego, empty of the illusion of self, through which can pour the world-sustaining Grace of God.

◈

XXIV

ROOT AND BRANCH

GOD IS THE UNIVERSE—but the universe is not God. If we take the natural world as an object which we worship as God, then we force her to submit to the human ego, thereby cutting her off from her transcendent Source. Consequently she sickens. But if we become transparent to the light of God Who transcends all manifestation, then nature will center around us, turn toward us, and, as it were, see God in us, thus allowing us to see God in her—a vision by which she is redeemed, purified and healed, first in perception and then in act. Only Humanity can attain the vision of a living universe, the robe of God, eternally streaming into manifestation from an Absolute Transcendent Source, the Heart of God. If we fail to reach this vision, we fail to assume our true human stature. If we fail to assume our true human stature, we become insects who devour the Earth: locusts with the faces of men (Revelations 9:7).

William Blake said: 'More! More! Is the cry of a mistaken soul; less than All can never satisfy Man.' As long as our perception is limited to space, time, matter and energy, we will never come into the field of this All; as long as we fail to intuit the sovereign reality of the eternal spiritual worlds, and so become capable of 'laying up our treasure in Heaven', we will be compelled to try and satisfy our God-given desire for His Infinite Reality within a finite earthly environment; we will squeeze the Earth dry in our thirst. This is why attachment to sense-experience is not the way to save the Earth, since it is precisely this which is destroying the Earth.

The roots of the environmental crisis are not in the transcendentalism of the Christian Middle Ages, but in the pseudo-immanentism, or glamorized materialism, of modern scientific myth, whose roots are in the neo-Pagan Greek revival of the Renaissance. Only if we know this material universe, reflection of God though it be, as only (in Blake's words) 'the hem of His garment' will we walk in balance on this fragile and uncompromising Earth.

❧

Even if you know that the world will end tomorrow, plant a tree.
—The Prophet Muhammad
(peace and blessings be upon him)

PART TWO:

ATLANTIS AND HYPERBOREA

Now I a fourfold vision see
And a fourfold vision is given to me
Tis fourfold in my supreme delight
And threefold in soft Beulas night
And twofold Always. May God us keep
From Single vision & Newton's sleep.

—William Blake, from a letter to Thomas Butts

PREFACE

As we have seen in *Part I*, the natural world, anchored on the material plane, can be a doorway to the imaginal world, the intermediary plane, the *alam al-mithal*. The natural world is composed of waves or cycles, some huge, some infinitesimal. And when seen from a standpoint outside it, a matrix that encompasses both its shape in space and its career in time, the natural world is revealed as a cycle in itself.

Not for nothing are many people who practice 'earth-based spirituality' also fascinated with ancient civilizations and past ages. The earth is not only a spacial form; she is also a temporal history. And at the point where we can see her form as united with her history in a single moment of space-time, we are also seeing the Imaginal Earth, the Eighth Clime, the Earth of Hurqalya.

The earth we know as a planet is also an aeon, and every aeon is a cycle of four ages, a Great Year. This is why an understanding of the 'cyclical mysteries', the inner meanings of the four seasons, the four directions and the four ages or *yugas*, can be another way of answering the question 'Who is the Earth?'

ATLANTIS
AND HYPERBOREA

AN INQUIRY INTO
THE CYCLICAL MYSTERIES

WITH A RECONSIDERATION OF RENÉ GUÉNON'S
RENDITION OF THE CYCLE OF MANIFESTATION
IN *TRADITIONAL FORMS AND COSMIC CYCLES*,
THE KING OF THE WORLD, AND *THE REIGN OF*
QUANTITY AND THE SIGNS OF THE TIMES

RENÉ GUÉNON AND HIS FOLLOWERS, notably Frithjof Schuon
and Martin Lings, take as one of their central cosmological principles that cosmic manifestation is entropic, not evolutionary: whatever has come into outer manifestation from the Unseen World has
already begun to die. Thus the traditional prophesies of the end of
'this world' are not the products of 'clairvoyance', arbitrary myth or
random visionary experience, nor are they projections or extrapola-
tions based on past events or present conditions. Rather, they are
based upon a cosmological context: the doctrine of the 'cycle of
manifestation', called by the Hindus the *Manvantara* or *Mahayuga*,
composed of four yugas or world-ages. Guénon and those influ-
enced by him consider the Hindu picture of this cycle to be the
most intelligible and complete, though a similar notion also
appears in Greco-Roman mythology, as well as among the Hopi
and the Lakota. (The idea of a cycle-of-manifestation may be found
the Abrahamic religions as well, but here the doctrine is less explicit,
more veiled in symbolism and allegory.)

The Greek word for such a cycle-of-manifestation is *aion*, which is translatable either as 'world' or as 'age'. When Jesus says 'behold I am with you always, even to the consummation of the world' or (according to a different translation) 'of the age', he is positing the reality of such a cycle. It is interesting that *aion* can be translated into English by either a spacial word ('world') or a temporal one ('age'). The reason for this is that *aion* denotes precisely a spacio-temporal reality. We visualize a year spatially as a cycle of four seasons, as we visualize a twelve-hour period in terms of the circular dial of a clock. Time is not purely linear; it is also cyclical, periodically returning to analogous (though not strictly identical) points: dawn, noon, dusk, midnight, the vernal equinox, the summer solstice, the autumnal equinox, the winter solstice. Anything that orbits, from the spinning of an electron around an atomic nucleus, to the wheeling of a galaxy, to (perhaps) the birth and death of the universe (a word that means 'one turn')—anything that exhibits *periodic motion*—is an example of a cycle in this sense. The word *aion* denotes the spacio-temporal reality of such a cycle—a cycle of time considered *sub specie aeternitatis* as a single quasi-spacial form, what the Hindus call the 'long body'. Any form or being that exists in time can also be viewed from the 'outside', from a point that is relatively eternal in relation to that form, and so seen as a single, 'simultaneous' history of itself, a kind of *histomap*. The term for this level of reality, in Eastern Orthodox theology, is *aeonian time*.

In any given cycle-of-manifestation, an eternal reality 'enters' time, moving from simultaneity toward succession. What is eternally present in synthetic mode is analyzed temporally, and therefore appears successively; in Plato's words, 'time is the moving image of Eternity.' But this move from eternity to time does not happen 'all at once'. Though eternity and time are in one sense absolutely discontinuous, necessitating a radical break, a 'fall' or 'ascension' in the passage from one to the other, in another sense the path from eternity to time moves through a number of stages, passing from the relatively more eternal toward the relatively more temporal, in the direction of the 'absolutely temporal'—a point which can never be reached, however, since 'pure sequence' would negate form absolutely, in which case there would be nothing to pass from one sequential phase to another. Manifestation is thus intrinsically entropic. An eternal form 'in the mind of God' appears in space and time, and is simultaneously veiled by its own manifestation. It becomes progressively more subject to history and contingency—and when it has consumed the energy of the initial impulse that brought it into manifest existence, it dissolves. Its dissolution unveils the eternal archetype of that form, which never entered manifestation—the naked radiance of which initiates the next cycle of manifestation.

It is in the context of this successive passage from eternity to time, or rather from aeonian to linear time, that the doctrine of the existence of earlier world-ages—which are not just earlier points in our own type of historical time—makes sense. And the *kind of sense* it makes also stretches from the relatively eternal to the relatively temporal. Near to the eternal end of the spectrum, a *yuga* (the *Satya-Yuga* or Golden Age) is symbolic; at the temporal end of the spectrum, a *yuga* (the *Kali-Yuga* or Iron Age) touches upon and embraces history as we understand it. Thus the true significance of the 'end of this world' cannot be grasped without an understanding of both the symbolic and the quasi-historical aspects of cyclical manifestation, which necessarily includes an understanding of the quality and meaning of the 'prior' states of the cycle, states which from one perspective are earlier in a temporal sense—given that we recognize that the quality of time was different in earlier ages—but according to another have 'priority' not in a historical sense, but in an ontological one.

Legends of ancient and mysterious lands, 'long ago and far away', legends of Agarttha, Shambhala, of the Terrestrial Paradise, the seat of Prester John, Atlantis, Lemuria, the Mount of the Prophets—stories like these always seem to collect around profound spiritualities, especially esoteric ones. On one level they are mere 'exoticism' or 'spiritual romanticism'. Those who entertain such dreams may never grow beyond them; they are in danger of letting their spiritual lives be trapped on the level of barren imagination.

But what of those who never allow themselves to entertain such dreams? Will, intelligence, sentiment are nothing without Grace—and one of the channels of Grace, at least in the initial stages of the

Path, may be the Imagination itself, which Blake called 'an Intellectual Fountain'. In apophatic contemplation (contemplation of God's Transcendence, based on the denial of His comparability to anything in the domain of manifestation), the profane imagination, based on individual fear and desire and its collective extensions, is negated; God is recognized as an unknowable Essence beyond all thought and feeling, beyond all name and form. But in cataphatic contemplation (the contemplation of God's Immanence, the recognition that He is in a sense comparable to all things, since without His Reality, no thing would be), Divine Imagination is born. Divine Imagination is objective Imagination, manifesting as the Imaginal Plane or *alam al-mithal*, the place where the 'image-exemplars' of Divine Realities appear as conscious, living symbols—as they do, on another level, in material reality.

So the mythopoetic lore of imaginal worlds, manifest on the psychic or intermediary plane, may have a valid and *spiritually operative* relationship to the world of metaphysical Principles, the intelligible plane—and this is definitely true of the 'cyclical mysteries', the legends of earlier aeons which were (and *are*) less constricted, less materialized than the world we presently inhabit. To project our contemporary concept of linear historical time backwards into earlier world ages is problematic, since different ages have different essential qualities; to consider previous *yugas* to be nothing more than earlier historical periods as we presently define them is to blind ourselves to these qualities. Yet earlier worlds are not mere allegories of higher ontological levels; they were (and are) real manifested worlds—formal worlds, not transformal intelligible Principles.

Traditional cosmology sees the present world and the present generation as 'descended' from the heroes and fathers of earlier ages who were ontologically more exalted than we are. They were taller than we are, lived for hundreds of years, were free from disease, etc. And these heroes, ancient kings, fathers, patriarchs or demigods were in turn descended from the gods, the celestial paradises, then from the intelligible Principles, and ultimately from the Creator Himself. God-as-Creator, in other words, was almost universally viewed in traditional cosmologies as the First Ancestor—literally

'God the Father'. And this hierarchicalization of history is also clearly discernible in the cyclical lore of many nations and religions—the Hindus, the Greco-Romans, the Mayans, the Hopi, the Lakota, the Australian Aborigines, and many African tribes. The 'earlier' a world-age is, according to these cosmologies, the more clearly it appears as eternal level of Being; the 'later' an age is, the more closely it resembles our idea of an historical period. So the 'trajectory' of a given cycle-of-manifestation is not a straight line, or even a circle, but rather a helix. Turning three times, it descends through four levels; and when it returns to its compass-point of origin for the third time—in other words, when it reaches its nadir—it undergoes a 'pole shift' from accelerating history to motionless simultaneity; the perspective changes from that of the last grains of sand speeding through the neck of the hourglass to the nearly motionless mass of sand below it, after which the glass is inverted: nadir becomes zenith. In Guénon's words, from *The Reign of Quantity and the Signs of the Times*, pp159–160:

It is sometimes said, doubtless without any understanding of the real reason, that today men live faster than in the past, and this is literally true. . . . If carried to an extreme limit the contraction of time would in the end reduce it to a single instant, and then duration would really have ceased to exist, for it is evident that there can no longer be any succession within the instant. Thus it is that 'time the devourer ends by devouring itself', in such a way that, at the 'end of the world', that is to say at the extreme limit of cyclical manifestation, 'there will be no more time'; this is also why it is said that 'death is the last being to die', for wherever there is no succession of any kind, death is no longer possible. As soon as succession has come to an end, or, in symbolic terms, 'the wheel has ceased to turn', all that exists cannot but be in perfect simultaneity; and this can also be expressed by saying that 'time has changed into space'. Thus a 'reversal' takes place at the last, to the disadvantage of time and the advantage of space: at the very moment when time seemed on the point of finally devouring space, space in its turn absorbs time. . . .

The zenith symbolizes the eternal (not temporal) point of origin, the Throne of the Most High God. And when zenith-as-eternal-origin is projected upon the horizontal plane, its representative compass-point is the North. (In the first chapter of Ezekiel, the vision of the Throne of God, which Sufis consider the apex of the created order, arrives from the North.) The *axis mundi*, the vertical path connecting Origin and manifestation, is vertical at any point on the earth's surface, as revealed by the fact that one of its central symbolic manifestations is the human spinal column and the erect stature of the human form. But for most of the human race, in whom the constant vision of the imaginal zenith is veiled, the pole of the earth considered as a psycho-physical object points to the North; therefore, when cyclical symbolism is expressed on the level of the four directions and the four seasons, its point of origin is North, and its moment of beginning the Winter Solstice. Among the mythic expressions of this fact is the legend of Hyperborea, the 'Land Behind the North Wind', the original homeland of the human race, a realm of eternal Spring. (The idea of a land of Eternal Spring in the far north was undoubtedly suggested by early explorers' tales of the arctic summer, during whose 'white nights' the sun never sets; this 'never-setting sun' was most probably the origin of the Hyperborean Apollo, one of whose epithets is *Sol Invictus*, 'The Sun Unconquered'.) The North Wind symbolizes the flow of cosmic manifestation, a flow that grows ever 'colder'—more contracted, more materialized, more literalized—as it departs from its celestial Source. The coldest point in this southward-flowing current of cosmic manifestation is the point of physical or spiritual death, which corresponds to the material North Pole; the growing warmth of the lands south of this Pole is organic, not spiritual, though it does reflect the higher warmth of Hyperborea itself, according to the law that the material plane furnishes the most complete and stable symbols of the Celestial one, though in inverted form. And above the material Pole is the Pole Star, the gateway to Hyperborea itself, which is *ontologically* higher than the material North as well as quasi-physically so, and which may indeed be compared to a Golden Age, the Paradise of Saturn, a land of Eternal Youth. To travel deliberately in the direction of the Arctic and its numbing cold suggests the path of apophatic contemplation, the realization of God in his purely Transcendent aspect, which requires a death to the world, an ascetical *metanoia* by which the contemplator turns his inner attention away from God's richly multiple and perilously confusing outer manifestation in the cosmic South, and toward 'the still point of the turning world' (in the words of T. S. Eliot from *The Four Quartets*). The rigor of this Transcendence is symbolized in Ezekiel 8:3 by the 'inner gate [of the Temple] that looketh toward the north, where was the seat of the image of jealousy, which provoketh to jealousy': the Transcendent God is a 'jealous God' who 'will have no strange gods before Him.' And this sacrificial turn is compensated for by a realization of the Terrestrial Paradise—not on some 'paradise island' in the South Seas, but in the subtle Imaginal Realm, the *alam al-mithal*. (The fact that Dante's Mount of Purgatory first appears as an island in the southern ocean, then undergoes an *enantiodromia* whereby the constellations of the polar North appear above it, symbolizes the withdrawal of the 'projection' of paradise from this world, which can never fully embody it, and the transplanting of that paradisical image to its true home on the Subtle Plane, in the realm that Muslims call 'The Earth of Hurqalya' or 'The Eighth Clime'.)

In the Old Testament the quarter of the South is represented by the Queen of Sheba, who was a pagan polytheist. In 1 Kings 10:1–10, she travels north to visit Solomon, brings him rich gifts, and blesses the One God; esoterically, this represents *apocatastasis*, the realization of the Divine Immanence, the restoration of all things in the Deity; the microcosmic aspect this restoration is the spiritual Path. Legend identifies the Shulamite of the Song of Songs with Sheba, who is in turn identified with the Queen of the South mentioned in Matthew 12:42 and Luke 11:31. The word *Shulamite* or *Shulamith* means 'daughter of peace', just as *Solomon*, whom legend identifies with the lover of the Shulamite in the Song of Songs (also called the Song of Solomon) may be translated as 'prince of peace'. While the Queen stays in the South, she remains hidden in self-involvement, coiled within herself, in a state the Hindus call *avidya-maya*, under which God is veiled by His own manifestation and the world of material nature appears sovereign and self-sufficient. When the

Queen of the South travels north, however, she is transformed from *avidya-maya* into *vidya-maya*—in Old Testament terms, from the 'foolish woman' to the figure of 'Wisdom' in the ninth chapter of Proverbs. She becomes the Shulamite of The Song of Songs, the bride of King Solomon, the King here representing the pole of Essence (*Purusha*) and the Queen the pole of Substance (*Prakriti*). While spiritual knowledge remains in the North it remains, in terms of human life, abstract and lifeless; and the South of the 'natural man' is all too literal, detail-oriented and *polytheistic* to embody it. But when the Queen of the South, the Shulamite, marries Solomon, the King of the North, then the Hyperborean Wisdom is fully realized and embodied in the 'minute particulars' of this world. She is his *shakti*, his *shekhina*, his principle of manifestation—and unless Wisdom is fully manifested, who can call himself wise? The marriage of North and South is the union of Heaven and Earth, the quality of which, as described in the text and commentary of hexagram 11 of the *I Ching*, Richard Wilhelm's translation, is precisely *peace*.

❡

In the mid–1970s, in San Francisco, California, I underwent a 'Hyperborean initiation', partly precipitated by the effects of my attending a Black Crown Ceremony presided over by the Gyalwa Karmapa of the Tibetan Karmapa Lineage of the Kargyupa Sect. This is one of the many public 'empowerments' provided by Vajrayana Buddhism, where the reflected grace of actual initiation is transmitted to the faithful as a kind of virtual initiation. During the course of the ceremony, the crowd meditated upon figure of the Karmapa, who doffed his pointed 'Tibetan mitre' and donned the Black Crown; this symbolized the dropping of his human identity and his assumption of the archetypal identity of the Bodhisattva Chenrezig or Avalokiteshvara. Those in attendance who were able to contemplatively 'follow' this transformation were given a foretaste of the transcendence of their own human limitations— something which could only have been be actualized through full initiation into the Vajrayana Way, with all its attendant transmitted graces and directed practices—or, perhaps, through a future initiation into an equally valid Way.

Some time after attending this ceremony, I had the following dream:

I am faced with a figure who is both the Gyalwa Karmapa and Merlin. He is wearing a conical hat like a 'sorcerer's' hat or a Tibetan mitre. He directs my attention with a pointing-stick to a large diagram: the Sun in the Moon Cradle, which is a down-ward-curving crescent Moon upholding the Sun as in a boat, with long wavy rays radiating from its disc. The figure points to six points on the lower crescent of the Moon and connects them, by means of lines, to six somewhat closer-together points on the upper arc of the Sun. I take this to mean that the Six Lokas or realms of illusion of the Buddhist Kalachakra (wheel of samsaric existence) are a projection into the relative world of Six Forms of Enlightened Mind.

Then the dream changes. I begin to dream about those ancient straight lines drawn on the earth's surface that John Michell, in his book *The View Over Atlantis*, calls 'leys'—except that in my dream they are called 'eber lines'. A voice says: 'An eber line can be drawn from any point to any other—but few remember that it is possible to draw an eber line straight up to God.'

This dream took place in 1976. I should make clear at this point that when I call this dream a 'Hyperborean initiation', I am using the word 'initiation' very loosely. Initiation is properly the establish-ment of a connection between a spiritual aspirant and an initiatory lineage—*silsila* in Sufi terminology—which stretches in an unbro-ken chain back to the founder of a given revealed religion. Initiation entails relationship with a spiritual master, and puts the aspirant in the field of the grace or *baraka* of the master and his predecessors, as well as giving him both the right and the power to practice cer-tain spiritual exercises. However, it is also true that particular visionary experiences can confer virtual or preliminary 'initiations' which, God willing, can sometimes function as more-or-less reli-able signposts on the Path—or rather, on the path to the Path. In times like ours, when legitimate spiritual authority is becoming harder to access, Divine Mercy sometimes provides valid guidance outside what we would think of as traditional channels. Yet unless such intimations finally result in the connection of the aspirant to an established traditional path, what might have been guidance will most often turn out to have been little more than delusion and a waste of spiritual potential—unless God wills otherwise.

The following 'exegesis' of the dream was composed in 1978:

Re 'leys': John Michell speaks of the discovery, made since the advent of aerial photography, of a series of absolutely straight lines of prehistoric vintage drawn all over the island of Britain. They are preserved in prehistoric earthworks, Roman roads, and short stretches of country roads and rustic paths. These lines take no account of geographic contours, but pass straight over all obstacles. Along a given ley (so named because the syllable 'ley' or 'lea' appears in many place-names along such routes) will appear a number of prehistoric sites, cathedrals—usually built on pre-Christian sacred points—or Celtic stone crosses, all in perfect alignment. On major sites, such as Stonehenge, Avebury, Woodhenge, and Glastonbury Tor, several leys are found to cross. Michell speaks of an identical system in China, where the lines are called 'dragon-paths'. Some believe these paths were channels established to conduct the flow of terrestrial magnetism, which was tapped on a seasonal basis by means of ceremonies conducted at sacred 'nodes' at various points along the track, points which were oriented to different heavenly bodies, as Stonehenge is toward the Sun. He shows that similar ceremonies took place in Britain, one purpose of which was to stimulate the fertility of the earth. The British leys, like their Chinese counterparts, are also mythologically associated with dragons. (Since Ireland is known to harbor no native serpents, could the legend of St. Patrick's driving of the snakes out of Ireland be a veiled reference to the suppression of such worship?) William Blake, too, hit upon the idea of leys, which in his system comprise a Druidic 'dragon temple' covering the whole of Britain.

So much then for 'ley'—but what about 'eber'? Very well: while thumbing through S. Foster Damon's *Blake Dictionary*, I ran into the name 'Eber', under the entry 'Peleg'. These names were taken from the genealogy in the tenth chapter of Genesis, where Peleg and Joktan are given as sons of Eber (the eponymous ancestor of the Hebrews), Peleg being so named because 'in his days was the earth divided'. And later, while reading Charles Squire's *Celtic Myth and Legend*, I came across the name Eber again: According to ancient Irish chronicles, Eber Scot was one of the sons of King Milé—leader of the Milesians, the first Celtic invaders of Ireland—who, in a dispute with his brothers over the new land, suggested and saw to it that the land was divided. (This tale can be explained, of course, as a monkish attempt to reconcile Biblical history with Irish legend.) So, apparently, leys *are* eber-lines.

Curious; and it would have remained little more than a curiosity, had I not found, 28 years later, in the year 2004, the following passage from René Guénon's *Traditional Forms and Cosmic Cycles*, p 24:

Hyperborea obviously corresponds to the North, and Atlantis to the West; and it is remarkable that although the very designations of these two regions are distinct, they may give rise to confusions since names of the same root were applied to both. In fact, one finds the root under diverse forms such as *hiber, iber* or *eber*, and also *ereb* by transposition of letters, signifying both the region of Winter, that is, the North ['Hibernia', for example], and the region of evening or the setting sun, that is, the West, and the peoples who inhabit both. . . . The very position of the Atlantean center on the East-West axis indicates its subordination with respect to the Hyperborean center, located on the North-South polar axis. Indeed, although in the complete system of the six directions of space the conjunction of these two axes forms what one can call a horizontal cross, the North-South axis must be regarded as relatively vertical with regard to the East-West axis, as we have explained elsewhere.

The eber-line which is 'drawn straight up to God' in my dream is precisely the axis mundi, the vertical path; the six points on the arc of the Moon connected with six other points on the arc of the Sun are the cosmic realities represented by the six directions of space, placed in relation to the formless archetypes of these realities within the Transcendent Intellect; they are possibly related to the Six Grandfathers in Black Elk's vision (see below), and to the vision recounted in the ninth chapter of Ezekiel of the 'six men [who] came forth from the higher gate, which lieth toward the north.' (See René Guénon, *The Symbolism of the Cross*; also Leo Schaya, *The Universal Meaning of the Kabbalah*, citations for 'serafim'.)

According to the religion of the Lakota, as recounted in *Black Elk Speaks* by John Neihardt, the North-South axis is called the Good Red Road, and the East-West axis the Black Road of difficulty. The point on the earth's surface where these two roads cross is *wakan*, holy—and they cross, of course, at the very point where one presently stands, wherever that may be. The Lakota call South 'the direction we always face'—and in saying this they speak as 'realized' Hyperboreans. Hyperborean spirituality, on the level of *aspiration*, is oriented toward the North, the 'original homeland' to which the aspirant wishes to return—but the Lakota, as it were, see themselves as occupying the North already, seated at the border between cosmic manifestation and the higher celestial worlds, at that point which the Qur'an calls 'the lote-tree of the farthest limit'. From this farthest point, this *ultima Thule* (cf. references to *Thule, Tula* in Guénon's *Symbols of Sacred Science, Traditional Forms and Cosmic Cycles*, and *The King of the World*), they gaze toward the South, into the visible universe. (On another level, the Lakota are oriented toward the East, which for ritual purposes they take as their sacred point; East is the point of Revelation, symbolized by the Eagle. The North, however, is the quarter of the Buffalo, the symbol of Totality; the coming of White Buffalo Cow Woman to the Lakota, to bring the sacred pipe, a symbol of the universe—analogous in many ways to the Hebrew Ark of the Covenant—was thus a Hyperborean theophany.) As Guénon suggests, the North-South axis, the Good Red Road, is the projection of the Vertical Path, the *axis mundi*—the Lakota symbol for which is the 'sacred tree within the hoop of the world'—upon the horizontal plane. If someone is an 'upright man'—in Hebrew, a *tzaddik*—then the Red Road, the path of life as indistinguishable from the path of the Spirit, stretches before him; his Way is clear. The Black Road, however, lies athwart this path, tempting him to deviate from his true course; this is the stroke that always 'thwarts' him. To one facing toward the South, the East-West axis stretches between Left and Right. To nod one's head 'yes' is thus to posit the *axis mundi*, and by extension the Red Road, while to shake one's head 'no' is to evoke the Black Road, the road of negation and self-contradiction.

In their annual migrations, the animal herds and the migrating birds travel the Red Road, along the magnetic lines of the earth. And insofar as the North-South axis is assimilated to the *axis mundi*, these migrations—especially those of the birds, most particularly the geese and the swans—naturally symbolize the passage between from this world to the next and back again. The Irish hear in the honking of migrating geese the voice of the 'Gabriel hounds', who carry the souls of the deceased to the other world. And certain Siberian shamans, who claim to have been 'born in the North', know how to ascend the *axis mundi*, symbolized by a birch tree, by riding on a goose, and so enter the other world. (Shamanism as a whole includes the clearest surviving examples of an ancient Hyperborean spirituality; see *Shamanism: Archaic Techniques of Ecstasy* by Mircea Eliade.) Likewise an epithet of certain Hindu yogis is *paramhamsa*, 'exalted gander' or 'gander of the beyond', undoubtedly indicating their ability to travel from this world to the next and back again, like the shamans. On a higher level, that of pure contemplation, this ability to cross the barrier separating life and death symbolizes the yogi's complete transcendence of conditional existence, by which 'this world' and the 'next world' are realized as one. And *hamsa* (by the Hindu principle of *nirukta*, a kind of etymology based on the morphological similarities of words rather than their historical derivation) is also *hang-sa*, the 'natural mantram' of the human breath. The assimilation of *hamsa* to *hang-sa* identifies the *axis mundi* with the human spinal column; as the geese fly north in the summer and south in the winter, so the breath rises from root to crown while inhaling (*hang*) and sinks from crown to root while exhaling (*sa*). By thus identifying his inhalations and exhalations with the seasons of the year, the yogi recognizes the cycle of the breath as a microcosmic *mahayuga*, an entire cycle of creation, dissolution and renewal. And since this cycle is experienced within the context of his own body, sitting erect and motionless in meditation, he experiences himself as no longer subject to the vicissitudes of cyclical manifestation, but as transcending them because encompassing them; he is thus effectively eternalized. The realized yogi, the *paramhamsa*, no longer turns within the cycle of the Great Year; the Great Year turns within him.

⑨

Perhaps René Guénon's most controversial book is *The King of the World*. Though just as rich in metaphysical insight and symbolic hermeneutics as the rest of his work, some have felt that Guénon had allowed himself in this work to be dazzled by a romantic exoticism of the Shangri-la variety, under the dubious influence of Ferdinand Ossendowski, whose stories of *Agarttha*, the mysterious and supreme spiritual powerhouse in Central Asia where the hidden King of the World reigns over the destinies of men, had seriously diverted him from his high calling of expositor of pure metaphysics. (Marco Pallis took pains to convincingly debunk Ossendowski's Asian travel tales in an article in *Studies in Comparative Religion*, Winter/Spring 1983.) It is clear, however, that Guénon was interested more in the cosmological symbolism of the King of the World and his hidden kingdom of Agarttha than he was in its literal reality—though he did not absolutely deny the possibility of a geographical Agarttha and a flesh-and-blood King of the World.

What are we to make of the myth of le Roi du Monde? And how real might this myth turn out to be in material, historical terms?

As myth, Guénon identifies the King of the World with the Hindu Manu—first lawgiver and archetypal human being—of the present cycle-of-manifestation, from whose name derives the very word 'man'. In one aspect, Manu is the representative in cosmic manifestation of God as the Primordial Ancestor—which is why it is in no way erroneous to identify him with the figure of Adam in the Judeo-Christian-Islamic tradition, especially since both Jews and Muslims recognize Adam as the First Prophet, and therefore—at least implicitly—as the Primordial Lawgiver. (In much the same sense, the Zoroastrian Gayomard is both First Man and First Prophet.)

In terms of the Hierarchy of Being, the meaning of The King of the World is as follows: Every material form simultaneously exists, in different modes, on every plane of being. A material stone, or plant, or animal, or human body is a truncated symbol, or partial reflection, of a form inhabiting the psychic or intermediate plane, which in turn symbolizes an entity residing on the angelic plane, which is the projection of an essence occupying the archangelic or intelligible plane, which itself is an emanation of the Logos, the transformal Origin of all form—the Logos also being God's eternal act of self-understanding in terms of cosmic creation viewed *sub specie aeternitatis*. If that stone did not simultaneously exist on all planes of being, if it were not in continuous, vital connection to the Logos, it could not appear in material reality: the Hierarchy of Being is the living 'stem' of every object in universal manifestation.

And what is true of stones is true of men. If the Human Archetype did not exist on all levels of being simultaneously, if he were not in fact the secret essence of the Logos itself, which the esoteric teachings of many traditions—Sufism within Islam, Kabbalah within Judaism, as well as various forms of Christian esoterism—identify with the Primordial Man, then there would be no men on earth. Therefore, no matter how far humanity has fallen, our Archetype, our *fitra* or primordial human nature, remains in its original integrity; this is 'The King of the World'. If every fall is, in one sense, a fall into illusion, then it must be true—in one sense—that the fall was illusory, that man never fell (though the consequences of this illusory fall are, unfortunately, all too real). Just as the Zoroastrian *fravashi* or *fravarti* is the aspect of my soul which never descended into material manifestation, so the aspect of Adam that never ate of the Tree of the Knowledge of Good and Evil is, precisely, le Roi du Monde. And those archaic religions, including many of the religions of Africa, certain aspects of Siberian Shamanism, and the Chinese/Taoist worship of the Yellow Emperor, which are oriented not to the Savior, the lawgiving Prophet or the redeeming Avatar—who comes later in the cycle in order to redress its corruption—but to the Original Ancestor, are worshipping the King of the World. Among these we may class the Mandaeans of Iraq, who worship not Christ, the Second Adam, but rather the Secret Adam, the First Man.

But in a much more restricted sense, we have already met, in various forms, the King of the World; he was alive in the 20th century and has survived into the 21st. Who else, after all, was the Gyalwa Karmapa when he donned the Black Crown? Who else is the Dalai Lama? Who else is the master of any Sufi order? Who else is the Shi'ites' Hidden Imam? There may or may not have been a single incarnate and universally-recognized King of the World in Central Asia, but all these well-known figures are certainly aspects, or instances, or delegates of that Kingship. The Jewish Kabbalists and Muslim Sufis both possess a lore of the Hidden Hierarchy and its Pole—that One whose degree of spiritual realization is pre-eminent in his own time, he whom the Sufis call the *qutb*, the Pole of the Age. In certain ages the *qutb* may be generally known, though his true identity as Pole of the Age will be not be realized by everyone. In other ages, he remains hidden. The master of every Sufi order is, in essence, the presence of this very *qutb* for his followers—if, that is, he is a true master. The Vajrayana Buddhists possess a similar lore. So in a certain sense, the King of the World is no mystery. Many today are outwardly familiar with him, though certainly not everyone who has heard of him recognizes him for who he is. So let us not wrangle too much about Guénon's own rendition of le Roi de Monde; it is clear that he was on the right track.

However, the lore of Vajrayana Buddhism does present us with a figure who, even more than the Gyalwa Karmapa or the Dalai Lama, seems to fulfill many of the criteria of Guénon's King of the World. If we replace the mysterious Kingdom of Agarttha (a word that means 'ungraspable', and thus may have originally have been more an epithet than a place name) with the much better attested Kingdom of Shambhala, we may find in the lineage of the Kings of Shambhala the possible prototype of Guénon's (and Ossendowski's) King of the World. Shambhala is a realm where myth and history intersect. As a geographical kingdom subject to terrestrial history, Shambhala may have been located north of the Tarim Basin in Central Asia—in eastern Turkestan, to be exact—which is north of Tibet; as a 'pure land', it is the 'area' of the *alam al-mithal* associated with the tantric tradition known as the Kalachakra, which forms an important part of Vajrayana Buddhism, though it is pre-Buddhist in origin and may have affinities with the Hindu Vedanta. The first of the Kings of Shambhala was Suchandra (878–876 BC?); the last will be Raudra Chakri (AD 2327–2427). *In The Wheel of Time: The Kalachakra in Context*, pp 56–57, by Geshe Lhundub Sopa, Roger Jackson, and John Newman, the King of Shambhala is described as follows:

> The Kalki (the lineage king) of Shambhala binds his hairlocks on top of his head; he wears a sacred headdress made out of dyed lion's hair and a crown marked with the symbols of the five Buddha families. He wears the costume of a universal emperor (*chakravartiraja*), and fortunate people are able to obtain the good path by simply seeing or touching him. The Kalki's emblematic earrings and the bracelets on his arms and legs are made of the gold from the Jambu River. The light of his ornaments mixes with the light that rises from the white and red luster of his body. It shines out to the horizon; it is so bright that even the gods cannot bear it.

> The Kalki has excellent ministers, generals and a great many queens; He has a bodyguard, elephants and elephant trainers, horses, chariots and palanquins. His own wealth and the wealth of his subjects, the power of his magic spells, the *nagas*, demons and goblins that serve him, the wealth offered to him by the centaurs, and the quality of his food are all such that even the lord of the gods cannot compete with him.

> Since the Kalki has a great many queens, he has many sons and daughters. However, when the Kalki-to-be is born (it does not matter whether he is the oldest son or not) there is a rain of white lotus flowers, and for one week prior to his birth the crown prince's body emits light like a radiant jewel. The queen mother, a daughter of one of the ninety-six satraps of Shambhala, is distinguished by the fact that at the time of her birth a rain of blue lotuses falls and a huge, previously unknown flower grows in front of her home. The Kalki and the queens possess the four aims of life [identical to the Hindu *ashramas*]; sensual pleasure, wealth, ethics and liberation. They never become sick or old, and although they always enjoy sensual pleasure, their virtue

never decreases. The Kalki does not have more than one or two heirs, but he has many daughters who are given as *vajra* ladies during the initiations held on the full moon of Caitra each year.

The fact that Shambhala is situated north of India, north of the Himalayas, north of Tibet and north of the Tarim River, and the related legend that the Kings of Shambhala counted among their servants the centaurs, emblem of the constellation of Sagittarius, whose month borders the Winter Solstice, mark it as a Hyperborean kingdom, the Polar King of which is the Kalki, around whom the entire universe of the Kalachakra revolves, just as the universe of China—or the entire universe seen as centered in China, identified as the 'Middle Kingdom'—revolved around the Emperor when he ascended the Altar of Earth within the Temple of Heaven in Beijing on the Winter Solstice, and worshipped the Pole Star. And the Kalki's 'heraldic' colors, red and white, are shared—interestingly enough—by a more familiar Hyperborean figure, Santa Claus. Like the Kalki, Santa is also served by elemental spirits, the elves. And the figure of Santa Claus and his reindeer has shamanic affinities as well. Some scholars associate the red and white costume of Santa Claus with the scarlet, white-spotted psychedelic mushroom *amanita muscaria* or fly agaric, which is used by certain Siberian shamans, and which mycologist R. Gordon Wasson considers to be the sacred *soma* plant mentioned in the Hindu Vedas—an attribution also accepted by Huston Smith. In the words of the Rig-Veda:

We've quaffed the Soma bright
And are immortal grown:
We've entered into light,
And all the gods have known.

Fly agaric is a favorite food of reindeer, which is why the drinking of reindeer urine as an intoxicant is (or was) practiced in Finland, and elsewhere in the far north. I hasten to add that the world-age when the use of such plant agents as aids to Enlightenment was possible without dire consequences, except in very rare instances, has obviously passed, as we can clearly see if we can view with sufficient objectivity the social and mass psychological effects of the use of 'psychedelics' or 'entheogens'. The 'psychedelic revolution' of the 1960s opened door of the mass psyche to everything imaginable, including the projection of traditional mystical lore of both the East and the West into the mind of the masses. This door, unfortunately, could never quite be closed again, and little has been coming through it for the past few decades but the influences of the elemental and the demonic—those 'infra-psychic forces' that Guénon, in *The Reign of Quantity*, saw as breaking into our realm through fissures in the 'Great Wall' separating the material and subtle domains, and ultimately leading to the dissolution of the present world. This is precisely the effect of psychedelics or 'psychic expanders' on the human soul: the attenuation, and sometimes the actual breaching, of the natural barrier designed to separate the human body, and thus the material plane itself, from the animic and psychic planes. What was spiritually possible in, say, the Silver Age or *Treta-Yuga*, is in no way possible in these last days of the Age of Iron.

So the Kalki of Shambhala would certainly seem to correspond in many ways to Guénon's Roi du Monde—though the question of whether or not the lineage of the Kings of Shambhala still remains hidden in Central Asia on the human, historical plane, or whether it has 'ascended into occultation' in the *alam al-mithal*, remains extremely difficult to answer.

The Pole is the lodestone to which all compass needles point, the point around which the Sun, Moon, and Stars revolve, the 'still point of the turning world'. The outer Pole is the extension of the earth's axis from the North Pole through and beyond the Pole Star. The inner Pole is the extension of the human spinal column from the *sahasrara* or crown *chakra* through and beyond the imaginal Zenith, as a vertical 'eber-line' or visionary ray of light. (The opening of the *sahasrara* by various tantric *sadhanas* is symbolized by the knot of hair that the tantric yogi—like the King of Shambhala—wears at the top of his head. And the blessing given by the Gyalwa Karmapa to the faithful after the Black Crown Ceremony is a closed fist planted firmly on the crown of the skull.) The secret kernel of the inner Pole is the point where the center of the human psyche is intersected by the ray of the Spirit; this is the site of the *qutb*, the Master of Masters. If the Heart is the Moon, the kernel of Spirit within it is the Sun—the indwelling transcendent Intellect—the Eye of the Heart. My dream of Hyperborean initiation was a visionary revelation of the Heart (the Moon) and the Spiritual Eye within it (the Sun.) (From another perspective, the Moon is the psychic aspect of the Heart, and the Sun the Spiritual aspect.) The word for 'heart' in Arabic is *qalb*, from the linguistic root QLB or QBL having to do with 'turning, overturning, turning around, returning.' The Moon, reflecting the Sun's light, turns through phases, but the Sun is constant. The spiritual Pole which appears within the darkness of the human psyche is the 'Light which lighteth every man that cometh into the world' [John 1:9]. It is the transcendent Light whose eternal moment is the Winter Solstice; it is the Midnight Sun. In the words of Henry Corbin from *Spiritual Body and Celestial Earth: From Mazdean Iran to Shi'ite Iran*, following the exposition of Shaykh Karim Khan Kirmani:

> The spiritual history of humanity since Adam is the cycle of prophesy following the cycle of cosmogony; but though the former follows in the train of the latter, it is in the nature of a reversion, a return and reascent to the pleroma . . . that is exactly what it means to 'see things in Hurqalya'. It means to see man and his world essentially in the vertical direction. The Orient-origin, which orients and magnetizes the return and reascent, is the celestial pole, the cosmic North, the 'emerald rock' at the summit of the cosmic mountain of Qaf, the very place where the world of Hurqalya begins. . . .[p71]

> The Earth of Light, the Terra Lucinda of Manichaeism, is also situated in the direction of the cosmic North. In the same way, according to the mystic 'Abd al-Karim Jili, the 'Earth of souls' is a region in the far North, the only one not to have been affected by the fall of Adam. It is the abode of the 'men of the Invisible', ruled by the mysterious prophet Khizr. A characteristic feature is that its light is that of the 'midnight sun', since the evening prayer is unknown there, dawn rising before the sun has set. [p72]

The Midnight Sun also appears in the first chapter of the Gospel of John, as 'the light [which] shineth in the darkness, and the darkness comprehendeth it not.' The psychic or natural man cannot encompass the Spirit or the reality of the Pneumatic Man; he must be darkened in spiritual self-annihilation before the Sun of the Spirit can dawn. And the only point where such annihilation can take place is the Spiritual Heart, which the Sufis identify as the *barzakh* or 'isthmus' between the 'two seas' of the material world and the realm of the Spirit, and which in mythological terms is considered to be the seat of the immortal prophet Khizr, the 'green one', and of the Earthly Paradise. Interestingly, the Arabic *barzakh* is quite similar in both sound and meaning to the Tibetan word *bardo*, which denotes either 'intermediary plane' or 'period of time between any two points considered as its beginning and end'—the letter 'z' often changing to 'd' according to the laws of linguistic transformation. The present moment is always intermediary between the world of material concerns and identifications and the realm of the Spirit; it is only Now, in what the Sufis call the *waqt*, that the annihilation of the natural or psychic man and the realization of the Pneumatic Man can take place. And only the Human Form, as epitomized by the Spiritual Heart as center and ruler of the psyche, stands as intermediary between the material and Spiritual worlds; it is this that makes us, in Muslim terms, *khalifa*, God's fully-empowered representative on earth.

☙

But was Hyperborea ever a terrestrial homeland? Geology shows us no sunken continent beneath the Arctic Ocean, which has led many to speculate that the North Pole once passed through Greenland, or some other point on the terrestrial globe. Yet a frozen wasteland, even if there were solid earth beneath it, is not a very hopeful candidate for the cradle of the human race—at least in terrestrial terms. Thus it is much more likely that Hyperborea refers to a *spiritual orientation* than to a geographical area. The Siberian shamans, the traditional Chinese, the Zoroastrians, the Sabaeans, and certain esoteric groups within Islam consider the North, not the East, or the West (sacred to the Greeks and the Irish, at least on one level) to be their spiritual point of orientation (or rather 'boreation'). 'Hyperboreans', then, are those who point to the Pole as their *celestial* homeland. Dante Aligheri, in his *Commedia*, reveals himself to be a Hyperborean in this sense. *Arktos*, the Greek word for 'bear', is the origin of our word *Arctic*, which is why the constellations circling the North Pole are called the Bears—and in the last cantos of Dante's *Purgatorio*, the Great and Little Bears appear above Dante's *Arcadian* Earthly Paradise at the summit of Mount Purgatory (which according to earlier cantos is supposed to be in the southern hemisphere!). In an outward sense, Mount Purgatory is in the South; in an inner one, it is in the North; the passage from the realm of the Great Mother Nature in the South to the Terrestrial Paradise of the North requires spiritual purgation, a 'pole shift' whereby the central attention of the Heart is shifted from outer manifestation to inner Source. (Hyperborea, however, may also have an historical, geographical significance; it may point to an actual northern culture-area dominated by shamanism, comprising Siberia and possibly Finland, and including, along with various other Arctic and North American peoples, the bear-worshipping Ainu of the Japanese northern island of Hokkaido.)

The Pole is the 'unwobbling pivot' (the Confucian term) around which the sky revolves, the still point of the turning world. If the 'turning world' is the Heart, then the Pole, the Qutb, is the Spirit. When the Heart starts to turn around the Pole instead of this or that object of desire or fear in the outer world, then it has returned from a state of dispersion (*tafriqah*) to a state of recollection (*jam*). This establishment of the Pole or *axis mundi* as the conscious center of the human microcosm is the aim of the practice of the Mevlevi Sufis known as 'turning', which has earned them the title of 'whirling dervishes'. The dervishes of the Mevlevi Order, founded by Jalaluddin Rumi, practice turning counterclockwise on the left foot, the foot whose extended vertical axis passes most directly through the human heart. To one turning counter-clockwise, the world appears to be revolving *clockwise* around him, just as the universe turns around the Pole Star; during the practice of such turning a 'pole shift' actually occurs when the sense that one is turning within a motionless world is suddenly replaced by the sense that one is actually motionless, that it is the world itself which is turning; one concretely experiences oneself as transmuted into 'the still point [or axis] of the turning world'. This, too, is a Hyperborean initiation, a virtual realization of the station of *qutb*—the *actual* realization of which, I hasten to add, is immensely more rare, difficult, and exalted. Rumi, who hailed from Afghanistan, may thus have been an agent for the introduction of certain elements of Central Asian, Hyperborean lore into the world of Sufism—which has a Hyperborean element in any case, as Henry Corbin has shown. The Sufi master Ruzbihan Baqli, to take one example, once dreamed that he had 'received oil from the constellation of the Little Bear.'

The *qibla*—another word which, like *qalb*, is based on the root QBL or QLB—is the direction in which Muslims face to pray. The outer *qibla* is toward Mecca. The inner *qibla* is oriented toward the Pole (in human terms, the Master), which is distantly symbolized by the North, more directly located at the Zenith, and quintessentially situated in the Heart, whose kernel—the 'Eye of the Heart'—is *al-Ruh*, the Spirit.

The North is the visible point of Eternity in the created order. The East is the quarter of the perennial renewal of creation and revelation, the land of the Rising Sun. The South is cosmic manifestation at its fullest degree of expansion and deployment, the point at which matter is most completely differentiated, established and infused by the Spirit, insofar as this is possible. It is the terrestrial world become so like Paradise that it is now possible to actually forget Paradise itself, with dire though sometimes long deferred consequences (like those Gauguin suffered when he 'went native' in Tahiti); so South is the quarter of the Great Mother, where God is most completely veiled by His own manifestation, and man most completely reconciled to, and engulfed by, the material world. And West is the quarter of matter in the process of becoming all-too-material, of the material world on its way to dissolution (cf. Guénon, *The Reign of Quantity and the Signs of the Times*, chapter 25, 'Fissures in the Great Wall'). The Spirit is deserting it; the Sun is setting; material manifestation is preparing to follow the Sun of the Spirit into the next *aion*, the next world. So West is the also the phase of death and the afterlife, the point at which Spirit sheds the husk of a dead world and begins its journey back to its original station in the heavenly world. This is why the Hesperides, the Fortunate Isles, the True West, all possess a certain nostalgic quality. This world has become too mate-rialized, too alienated from the Spirit; the Golden Age, the Silver Age, even the Bronze Age are past; the path toward restoration now leads through the Gates of Death, to a land 'beyond the sunset'. Like-wise the Western Paradise of Amitabha Buddha, the 'pure land' of Shin Buddhism, is the place where those who have not achieved per-fect total enlightenment in this world can, by the grace of Amitabha, 'ripen' into enlightenment after death.

The legend of Atlantis, the lost continent of the West, has this nostalgic quality—a quality accompanied (and also partly veiled) in the subconscious memory of the human race by the shock of the cataclysm that destroyed it. This quality is best rendered, perhaps, by the spine-chilling indigenous music of the Andes (played on flute, pan-pipes and *charango*, the Andean mandolin or treble gui-tar whose sounding-box is an armadillo shell), and in another way by the heart-rending nostalgia of certain Irish and Scottish aires. The ship of human memory, riding the waves of such music, travels back 30,000 years, to human worlds now sunk in the collective unconscious by the shock of global cataclysm. All is lost—and yet, for those willing to make the ultimate sacrifice, all may be found again, fresh and incorrupt, bursting with energy, in the Land of the Ever-Young.

This is the psychic and cosmological quality of the Atlantis leg-end. But could it also have a historical aspect? Orthodox geology says 'no'. Whatever we might fantasize about the Canaries, the Azores, the West Indies, there is simply no geographical evidence of a sunken continent anywhere in the Atlantic. And yet there are cer-tain scholars who make a very good case for the material, historical existence of Atlantis—simply by identifying Atlantis with North America, or the Americas as a whole. The Aztecs, we should remember, who are thought to have invaded and conquered the Toltec Empire of Mexico from a point of origin somewhere in the territory now claimed by the United States, named their former homeland as *Aztlán*—a word close enough to *Atlantis* to make one's hair stand on end.

So according to this theory, I am in Atlantis now. But the conti-nent I inhabit is certainly not sunken—unless we admit that it is sunk in materialism, overwhelmed (in Blake's words) by 'the sea of Space and Time'. So whence comes the legend of the *lost* Atlantis, perhaps symbolized in Greek legend by the runner Atalanta, the woman no man could catch? A sunken continent may legitimately be compared to a woman who has forever denied her lovers any possible access to her—and what man can outrace the setting Sun? The men who raced Atalanta to win her hand, and lost, also lost their lives—this being the precise quality of the Western Quarter, the land of 'futurism', where time accelerates and form is destroyed (cf. *The Reign of Quantity and the Signs of the Times*, p159ff). And in line with Guénon's assertion that Hyperborean terms were later applied to Atlantis, one of the epithets of Atalanta is *Arcadian*. When she finally was outraced by her future husband Hippomenes, it was through the agency of three *golden apples* given him by Aphrodite from her own temple precincts in Cyprus, the last of which Atalanta stooped to pick up when Hippomenes threw it, thus losing her stride. Golden apples immediately suggest the apples of the Hesperides, the Western Isles—and though the island of Cyprus is in the eastern Mediterranean, it is certainly west of the continental Near East. The virgin Atalanta was able to inspire in men a nostalgia for the lost Paradise *without feeling it herself*—until, that is, she touched the third apple. Perhaps we can glimpse in this legend one aspect of the Western Mysteries: if the Hesperides can be made to feel nostalgia for man as well as man for the Hesperides, then the Western Paradise may be won. In the words of William Blake, from his Introduction to *Songs of Experience*: 'O Earth O Earth return/ .... Why wilt thou turn away/The starry floor, the watry shore/Is giv'n thee till the break of day.'

So when, and how, was Atlantis lost? A.G. Galanopoulos and E. Bacon, in *Atlantis: The Truth behind the Legend* (1969), J.V. Luce, in *The End of Atlantis: New Light on an Old Legend* (1969), and Charles Pellegrino, in *Unearthing Atlantis* (1991), theorize that Atlantis was actually the island of Thera or Santorini, situated—like Cyprus—west of the Mediterranean coast of the Holy Land, Thera being directly north of Crete. It is a volcanic island which, in 1450–1500 BC (some date the event c.1628) violently exploded when its erupting volcano split at the side, allowing an inrush of sea water.

The explosion was several times larger than that of Krakatoa, the most powerful volcanic event in recorded history, which was also destroyed in a steam explosion. This cataclysm devastated the Mediterranean coasts, sent a towering tsunami crashing over the island of Crete, darkened the sun with volcanic ash, and effectively destroyed the Cretan/Minoan/Mycenean maritime civilization. It began a series of migrations and wars, one of which was the invasion of the Greek peninsula by the Doric tribes, the ancestors of the 'classical' Greeks. Some scholars also theorize that the ten plagues (or some of them) which preceded the exodus of the Hebrews from Egypt were actually volcanic in origin: the hail mixed with fire, the turning of the Nile to blood along with the death of all the fish, the darkness which covered the land, can all be put down to the effects of volcanic cinders and ash. And the parting of the Red Sea, which later closed over the Pharaoh's army, suggests the arrival of a tsunami, during which the sea-level first sinks and then catastrophically rises; such a tsunami would have been possible (or rather inevitable) if—as some think—Sinai was at that time a strait rather than an isthmus; it would certainly have been more feasible for the Children of Israel to have a crossed a narrow strait rather than the Red Sea as we know it today. And the 'pillar of cloud by day and pillar of fire by night' that the Hebrews followed through the wilderness is a fair description of a rising volcanic plume.

Our major source for the Atlantis legend are the *Critias* and *Timaeus* of Plato, who recounts a history of the lost island supposedly based on an account that Solon heard from the priests of Egypt. Plato's description of Atlantis as an island of concentric rings of land and water corresponds in some ways to the geology of Thera; and the legend that Atlantis was situated beyond The Pillars of Hercules—the Straits of Gibraltar—is possibly explained by the fact that Thera is actually west of another formation, in the eastern Mediterranean, which is also named The Pillars of Hercules.

But what of the *American* Atlantis mentioned above? Ivar Zapp and George Erikson, authors of *Atlantis in America* (1998), maintain that 'Atlantis' sank beneath the waves when, around 12,000 years ago, sea levels abruptly rose due to melting polar ice, thus inundating coastal America. The authors give evidence to support their contention that before that time America was host to an advanced maritime civilization capable of crossing the Atlantic. This theory is further supported by the fact that certain *metis* societies (inter-tribal medicine societies) among the Native Americans of North America claim that they were in contact with Europe in ancient times. Travel across the Atlantic was dangerous; few probably attempted it, but some likely did. Regular trade routes might or might not have been established, but holders and seekers of spiritual lore and technical expertise may well have attempted the journey, given that knowledge is weightless, and takes up no space. Various Greek philosophers visited Egypt, Persia and reputedly India in ancient times; Bodhidharma took Buddhism from India to China; sages, as well as craftsmen and artists (including poets and musicians) were among the most traveled groups in earlier times, long before Marco Polo. Everybody else might be hunkered down, but the craftsmen and the sages were abroad, restlessly searching, or bursting with a knowledge that commanded them to spread it broadcast, like wind-blown seeds.

Can the Mediterranean and American Atlantises in any way be reconciled? Some legends of Atlantis speak of two Atlantises, an earlier and a later one. Zapp and Erikson's submerged coastal America, then, might correspond to the earlier Atlantis, perhaps also recalled by the legend of Noah's flood, and Thera to the later one, which might possibly be the origin of certain events recounted in Exodus. After the 900 years separating Plato from the most common date given for the destruction of the Greek island, certain legendary material about the earlier Atlantis could well have become attached to the story of the destruction of the later one; the characterization 'island continent' may in fact be the product of a confusion between the submergence of part of a continent and the destruction of an island.

The submergence of coastal America would have been either gradual or cataclysmic. A slow melt of polar ice would not have destroyed the Atlantean civilization—unless it forced the coast-dwellers back into an interior occupied by hostile and militarily superior nations. They would always have had a coast, and time to move any cities inland. A fast melt would correspond more closely to the Atlantis legend as we know it. And if trans-Atlantic trade, however sporadic, had existed, its sudden disappearance would indeed have suggested—and actually represented—the destruction of a world, especially if the traders hailed from a civilization that was either spiritually higher or technologically more advanced than was the Old World in that age. The voyages of Thor Hyderdahl across much of the Pacific on a balsa raft, and across the Atlantic in a reed-boat constructed according to an Egyptian pattern—both in order to demonstrate that ancient legends of epic sea-journeys might have had some basis in fact—as well as the maritime exploits of the Polynesians and others, testify to the likelihood that world travel was much more common in the archaic world than we once believed. If a reed boat could cross the Atlantic (reminding us of the legend that, after the destruction of their former world, the Hopis arrived in the New World *inside floating reeds*), then it could certainly have been crossed by the much more technologically advanced ships of the Minoan civilization.

We are used to seeing the Mediterranean largely as a 'closed sea' until the Vikings, and later the Renaissance explorers, opened the mind of Europe to the Atlantic and the New World. But the maritime technology that would have allowed Europeans to cross the Atlantic had been available since the Roman Empire, and even before that. Why (outside of the Roman colonization of Britain) was it never used? It is possible to speculate that the shock of the submergence of coastal America by melting ice, which would certainly have also submerged much of the coast of the Mediterranean, as well as the lands called *Lyonesse* in British legend—followed in later centuries by the destruction of Thera, which liquidated in one stroke the most advanced maritime civilization the Old World had produced up to that time—created a sort of collective taboo in the European psyche against sea-travel beyond the pillars of Hercules, and possibly against expansive maritime imperialism in general, which would have been viewed as actions likely to anger the gods. This taboo was effectively broken by the Vikings, relative newcomers in Western Europe, whose historical memory stretched back not to the archaic civilizations of the Mediterranean and Near East, but towards the heartlands of Asia. Furthermore, the opening of the

Atlantic and the New World to exploration during the Renaissance may have awakened long-buried memories of the Western Atlantis in the form of fantastic and legendary goals sought by some of the explorers and conquistadores: the Seven Cities of Cibola, and especially the Fountain of Youth, which clearly corresponds to the fountain of the water of life—or the water of creative manifestation—situated by Dante at the summit of Mount Purgatory, in the Terrestrial Paradise. (The taboo against 'westering' appears in the 'Atlantean' Canto 26 of Dante's *Inferno*.)

But history, as we know it, can never be the history of Paradise; and even less can future history fill this role. Those who wish to regain the lost Paradise through explorations of the West—the world of conditional, historical life, of evolutionism and scientism, the point where Spirit is finally reduced to matter, and where matter, seemingly bereft of Spirit, proceeds to its inevitable dissolution—can only reach that distant Land through the gates of death. For them, Paradise is nowhere else but beyond the grave. The true spiritual Path, however, the Path of 'death-in-life', is *counter-clockwise*, moving against the *clockwise* current of Nature, the flow of manifestation; it is effected by a *reversal* of the cosmogonic process.

The four *yugas*, the four seasons and the four directions are analogous. The *Satya-Yuga* or Golden Age or Hyperborean Paradise is in the North; its season is Winter. The *Treta-Yuga* or Silver Age is in the East; it's season is Spring—not the eternal Spring of Hyperborea, surrounded and protected by the rigors of Winter, but the temporal Spring of eternal renewal. (When Adam and Eve were exiled from their Hyperborean Eden—Hyperborean by virtue of the Tree of Life, which is the *axis mundi*—and traveled 'to the east of Eden', this represented the transition from *Satya-Yuga* to *Treta-Yuga*.) The *Dvipara-Yuga* is South and Summer; the *Kali-Yuga* is West and Autumn. The *Satya-Yuga*, according to Hindu doctrine, is four times as long as the *Kali-Yuga*, the *Treta-Yuga* three times as long, and the *Dvipara-Yuga* twice as long: as we have already seen above, time speeds up as the cycle descends, and is progressively changed from a cyclical to a linear form. The cycle of the four *yugas* is thus a clockwise-rotating path from Transcendent Source to visible manifestation (North, East, South, West), and is always centrifugal. As that Source, or the memory of It, becomes more and more externalized and literalized, the lost Paradise is progressively, uselessly, obsessively, ironically, and disastrously sought in physical exploration, revolutionary social experiments, fantastic 'evolutionary' hopes, technological 'progress', and finally in the deconstruction of the human form itself. This is the way of collective destruction and the end of the present world: the 'West' is always 'Atlantis'. But the counter-clockwise-rotating Path of Return-to-Source (West, South, East, North) is centripetal, given that North, the pivot of the heavens as seen from the Northern Hemisphere, is the primal Center. The Spiritual Path is a motion away from literalism, from ever-accelerating linear time, from the 'nightmare of history'. It is a Path whereby the image of Paradise is withdrawn from projection on the outer world, and re-deployed on an ascending scale of ever-higher planes of Being, each more radically 'inner' than the preceding one. Instead of falling from the Eternal North to the East of perpetual renewal, thence to the South of immersion in material conditions, and finally to the West of chaos and dissolution, the Way demands that chaos (the West) be overcome through stabilization of one's bodily and 'practical' material life (in the South); then that one's life become open to the waves of spiritual insight, vitality and perpetual renewal emanating from the East; and finally that one avail oneself of this current of Divine Grace, like a salmon traveling upstream to spawn, till one finally arrives at the threshold of the *Qutb*, the Eternal North, the Gateway that leads beyond the cycles of nature, the still Point of the turning world. Once this Gateway is passed and the eternal Center realized, the universe is transformed from an obscuring veil into the universal Theophany; the natural world from which one had to withdraw one's inner attention in order to travel the spiritual Path is restored on a higher level; rigorous Transcendence is transformed into merciful Immanence. One may now *look back* on the world of nature, and see it as circumambulating, in perfect obedience and punctuated and rhythmic praise, that adamantine Point—a Point which, speaking now in anthropological rather than cosmological terms, is the *atman*, the eternal Witness, the Eye of the Heart. And note: while the centrifugal path of manifestation is collective, and only becomes more collective as the cycle winds down, the centripetal

Path of Return is always individual, and becomes ever more so as the transcendent *atman* is approached. As the Gospel parable teaches us, the grace of God abandons the ninety-nine and seeks the One. And when those many 'ones' meet in the Eternal Human Form—when they pass, together and alone, beyond the dimensions of space and time—this convergence of irreducible solitudes as One Presence is the reality of the *apocatastasis*, the restoration of all things in God.

Atlantis is Memory—the Western gate to the other world, whose central sign and agent is natural death; this perhaps explains the addiction of the 'Atlantean' Toltecs, Aztecs and Mayans to the practice of human sacrifice. It is *pitri-yana*, the Way of the Ancestors, who, by the power of cosmic nostalgia, become their own children again, forever, in endless repetition: 'the sun also ariseth, and the sun goeth down, and hasteth to his place where he arose' (Ecclesiastes 1:5). Hyperborea is the rigor of Transcendence, the Northern gate that leads to the other world by the power of sacrifice and spiritual ascent, rising bodily out of the South, then turning back to contemplate universal manifestation from the eternal Point beyond it. Its emblem is not *literal* human sacrifice, but rather the spiritual death and resurrection of the shaman, or of the esoteric initiate. It is the *deva-yana*, the Way of the Gods; 'I call it death-in-life and life-in-death' (W.B. Yeats, from *Byzantium*).

The Eternal Gates' terrific porter lifted the Northern Bar. . . .

—William Blake, *The Book of Thel*

AFTERWORD

THE HYPERBOREAN TEMPTATION

AS THE PRESENT CYCLE OF MANIFESTATION draws to its close, it becomes increasingly possible to discern the shapes of earlier world-ages; the faculty of cosmic intuition, as well as the sciences of history, archaeology and geophysics, respond to the 'thinning' of the cosmic environment. Because the space-time barriers of the present world are becoming ever more transparent as the cycle nears its end, the lineaments of the Golden Age of Hyperborea become progressively easier to make out through those fading outlines. Yet we still occupy the *Kali-Yuga*, and will until 'the consummation of the age'. So it becomes necessary to ask the following question: Exactly what place, if any, do 'Hyperborean intuitions' occupy in the spiritual life? Do they empower the spiritual Path, or merely distract one from it?

It is perhaps easier to see the dark side of the 'Hyperborean Renaissance' than its spiritually effective aspect. The present flooding of the western world with Hyperborean psychic material—as the west was flooded by 'Eastern' material during the 1960s—is happening on a deeply unconscious level. When the Western masses were exposed to the lore of Hinduism and Buddhism, more-or-less legitimate teachers and exemplars also arrived who could explain it to them, and guide them by means of it. But no viable and unbroken lineages exist which lead back to the Hyperborean Age; shamanic spirituality, which is our closest approach to such transmission, is largely degenerate, and in any case does not originate from the Golden Age proper, but from a later *yuga*—possibly the Silver Age—when the work of rebalancing the cosmic environment and fighting off the incursion of demonic forces had already replaced the 'mass theophanic consciousness' of the *Satya-Yuga*. And in the absence of true lineages and teachers, the present Hyperborean incursion finds among its available interpreters only individual 'geniuses', as well as cranks, wizards and occult fantasts. René Guénon and his followers are almost alone in providing an adequate doctrine of the Hyperborean Age, and even in their case, such doctrine cannot form the *upaya* (method) of a true spiritual Path. We can see in the Goth culture of today's youth—which does indeed bear certain clear affinities to Nazism and the doctrines of Friederich Nietzsche—the necessary effect of the dawning of Hyperborean influences within the shrunken confines of the *Kali-Yuga*, which must inevitably result in such influences being taken on much too low a level.

The Six *Lokas* of samsaric existence, in the system of the Kalachakra, are as follows: the humans, the animals, the gods (*devas*), the hell-dwellers, the warrior-demons (*asuras*) and the hungry ghosts (*pretas*). If we assign these modes of existence to the six directions of space, the following attributions seem plausible: *Devas* at the Zenith (Heaven); hell-dwellers at the Nadir (the Pit); animals in the South (the natural world); hungry ghosts in the West (the quarter where life and reality are most radically depleted); the human world in the East (the place where incarnate existence is open to spiritual illumination); and the *asuras* or warrior-demons in the North. (This last attribution is supported by the fact that the *nagas* and *rakshasas* [goblins] who serve the Kalki of Shambhala are two classes of *asuras*.) The *asuras*, like the Norse giants and the Greek titans, are always trying to 'take Heaven (the world of the *devas*) by storm', but they never ultimately succeed. And it is certainly not hard to see the 'warrior-demon' aspect of the 'Hyperborean' Goth culture, since it is composed of little else. Just as the North is the projection of the Zenith onto the horizontal plane, so the 'spirituality' of the North is necessarily on a lower level than that of the Heaven of the Zenith, the doorway to the celestial worlds. Transposed into terms of the spiritual Path, this indicates a 'prometheanism' by which the Path is reduced to a destructive and self-defeating exercise of titanic self-will, which is the ego's perennial misinterpretation and mis-application of spiritual zeal. Our road of access to God is only through our need; our point of contact with the Absolute is only through our self-annihilation. The *asuras* and their worshippers, however, will have none of this. Like Nietzsche, they scorn humility and self-annihilation as mere bourgeois cowardice and sentimentality; they fail to understand these virtues as in fact representing an unsentimental courage of the highest order. The Greater Jihad, the 'war against the self', is infinitely hotter and more rigorous than the Lesser Jihad against outer conditions; if it were not for the outpouring of God's Mercy, none would survive it, all would be defeated. How many people (dare I include among them the warriors of al-Qaeda?) flee into outer conflicts of all kinds, and ultimately into the arms of self-destruction, so as to avoid encountering the terrifying face of the Ego, the *nafs al-ammara*, the Enemy Within? Suicide—especially when sacrilegiously identified with service to God—is infinitely easier to contemplate than this grim and terrible meeting.

Mythologically speaking, the *asuras* are self-willed titans perpetually fighting to conquer Heaven; in terms of the spiritual Path, they are Heaven's *gargoyles* or *temple-guardians*—that mass of hindrances, produced by our *own* self-will, which prevents us from attaining the celestial realms. (The same reality is represented, in Genesis 3:24, by the 'cherubim, and a flaming sword which turned every way' that God placed on the border of Eden after the Fall to guard the Tree of Life.) But it should also be noted that Buddhism does not take the *devas* to represent Heaven in our sense of the word. The Kalachakra Tantra characterizes the 'gods' as complacent and ignorant; their world, though pleasant, is just as much a product of anger, lust and ignorance as the other five *lokas*, and is equally impermanent. Even the world of the gods is part of the 'round of existence' from which sentient beings must be liberated.

But the North as a lower, psychic / cosmological reflection of the Zenith, not only tempts to promethean self-will, but to an even more dangerous inflation of the intuitive/intellectual faculty of the human soul—these two transgressions together comprising the sin of spiritual pride. From the standpoint of the Imaginal Pole, it seems as if all manifestations of spirituality can be seen, evaluated, and judged. And on a certain level this is true. Yet such a Hyperborean vantage-point, which alone makes possible an understanding of the

Transcendent Unity of Religions (to use Frithjof Schuon's term), is not the true *ultima Thule*. The perspective it provides is necessary to any deep understanding of religious forms in these times, an understanding which becomes progressively more crucial as the *Kali-Yuga* winds down. But in terms of the concrete practice of the spiritual Path, though it may be the source of a valuable sense of context, it is neither effective doctrine nor useful method, neither *prajña* nor *upaya*. Hyperborean knowledge in effect makes us too 'big' to pass through what Jesus called 'the eye of a needle'. We come to such knowledge partly through identification with it, but such identification must be broken before we can make real spiritual progress. Hyperborean intuition is what the Sufis call a state (*hal*), considered as a gift of God; it can never be a station (*maqam*), defined as a fruit of the realization of the virtues, and of the prime virtue of the spiritual Path, self-effacement. It is a knowledge which may, in the mystery of God's will, have the power to re-orient us to the eschatological grace of the latter days; but as a *possession*, it is useless. In terms of the work God demands of us, it does not represent even a single step.

The memory of the Golden Age such as is becoming increasingly possible in these apocalyptic times will necessarily suggest, to some, that we are about to return to such an age; this is perhaps the fundamental error of the New Age movement. But the truth is, we cannot, on a collective level, return to the Golden Age of this cycle, nor can the human collective reach the Golden Age of the cycle to come except through the door of physical death. To imagine that a human race sunk in the *Kali-Yuga* can collectively attain to the exaltation of the *Satya-Yuga—in historical terms—*is, in fact, one of the pillars of the coming Kingdom of Antichrist, and of the globalization process which is its herald; when the Golden Age is seen and sought through the eyes and hands of the Age of Iron—as with the 'millennium' of Adolph Hitler—the result can only be the Great Inversion that René Guénon spoke of in the 39th chapter of *The Reign of Quantity*. (Interestingly, tradition teaches that the Antichrist will spring from the Tribe of Dan, who occupied the northernmost region of Judea in biblical times.)

Hyperborea is too big for us. In the vast majority of cases, the faculty that may intuit it can only reach that exalted place at the expense of the other faculties of the soul. When the spiritual intuition is inflated, the will is depleted or perverted, the affections poisoned, dissipated, or frozen solid. We must pay for such knowledge, and the only avenue by which such payment can be made is the spiritual Path. Only the Path of self-annihilation can transform the Hyperborean inflation from a potential demonic incursion into a *felix culpa.*

Hyperborean or Primordial spirituality—outside of a few very rare cases (and God knows best)—cannot be a Path for us. For the vast majority of believers, and also the vast majority of *esoterics,* only the revealed religions can provide spiritually efficacious forms of the Path, forms which are addressed to the fragmentation, dissipation, petrification, and general *smallness-of-soul* of modern and postmodern humanity. We cannot overcome this smallness by inflating ourselves (given that we can't inflate ourselves without simultaneously dwarfing ourselves); we can only put our shrunken condition to good use by transforming it into self-effacement, humility and trust in God. To rival God by allowing our souls (in Sufi terms) to be 'qualified by the Names of Lordship' is to lose Him; to approach him in our essential poverty—as when the Lakota brave will 'lament' for a vision from Wakan Tanka—is the only way to encounter the riches of His Mercy, His Knowledge, and His Power.

๑

The Zahir is the shadow of the Rose
And the rending of the veil.

— Attar